Mazhar Ali

Manipulating sudden death of quantum entanglement

Mazhar Ali

Manipulating sudden death of quantum entanglement

Quantum Control of Entanglement Sudden Death for Qubit-Qubit and Qubit-Qutrit Systems

Südwestdeutscher Verlag für Hochschulschriften

Impressum/Imprint (nur für Deutschland/ only for Germany)
Bibliografische Information der Deutschen Nationalbibliothek: Die Deutsche Nationalbibliothek
verzeichnet diese Publikation in der Deutschen Nationalbibliografie; detaillierte bibliografische
Daten sind im Internet über http://dnb.d-nb.de abrufbar.
Alle in diesem Buch genannten Marken und Produktnamen unterliegen warenzeichen-, marken-
oder patentrechtlichem Schutz bzw. sind Warenzeichen oder eingetragene Warenzeichen der
jeweiligen Inhaber. Die Wiedergabe von Marken, Produktnamen, Gebrauchsnamen,
Handelsnamen, Warenbezeichnungen u.s.w. in diesem Werk berechtigt auch ohne besondere
Kennzeichnung nicht zu der Annahme, dass solche Namen im Sinne der Warenzeichen- und
Markenschutzgesetzgebung als frei zu betrachten wären und daher von jedermann benutzt
werden dürften.

Verlag: Südwestdeutscher Verlag für Hochschulschriften Aktiengesellschaft & Co. KG
Dudweiler Landstr. 99, 66123 Saarbrücken, Deutschland
Telefon +49 681 37 20 271-1, Telefax +49 681 37 20 271-0, Email: info@svh-verlag.de
Zugl.: Darmstadt, TU Darmstadt, Diss., 2009.

Herstellung in Deutschland:
Schaltungsdienst Lange o.H.G., Berlin
Books on Demand GmbH, Norderstedt
Reha GmbH, Saarbrücken
Amazon Distribution GmbH, Leipzig
ISBN: 978-3-8381-0783-7

Imprint (only for USA, GB)
Bibliographic information published by the Deutsche Nationalbibliothek: The Deutsche
Nationalbibliothek lists this publication in the Deutsche Nationalbibliografie; detailed
bibliographic data are available in the Internet at http://dnb.d-nb.de.
Any brand names and product names mentioned in this book are subject to trademark, brand
or patent protection and are trademarks or registered trademarks of their respective holders.
The use of brand names, product names, common names, trade names, product descriptions
etc. even without a particular marking in this works is in no way to be construed to mean that
such names may be regarded as unrestricted in respect of trademark and brand protection
legislation and could thus be used by anyone.

Publisher:
Südwestdeutscher Verlag für Hochschulschriften Aktiengesellschaft & Co. KG
Dudweiler Landstr. 99, 66123 Saarbrücken, Germany
Phone +49 681 37 20 271-1, Fax +49 681 37 20 271-0, Email: info@svh-verlag.de

Copyright © 2009 by the author and Südwestdeutscher Verlag für Hochschulschriften
Aktiengesellschaft & Co. KG and licensors
All rights reserved. Saarbrücken 2009

Printed in the U.S.A.
Printed in the U.K. by (see last page)
ISBN: 978-3-8381-0783-7

Dedicated to the loving memory of my mother (Ammi):
Pari Jaan Jhanghiri.
&
To my dearest brother and wellwisher (Bhai jaan):
Dr. Liaqat Ali.

Abstract

This thesis is a theoretical study of entanglement dynamics and its control of qubit-qubit and qubit-qutrit systems. In particular, we focus on the decay of entanglement of quantum states interacting with dissipative environments. Qubit-qubit entanglement may vanish suddenly while interacting with statistically independent vacuum reservoirs. Such finite-time disentanglement is called *sudden death of entanglement* (ESD). We investigate entanglement sudden death of qubit-qubit and qubit-qutrit systems interacting with statistically independent reservoirs at zero- and finite-temperature. It is shown that for zero-temperature reservoirs, some entangled states exhibit sudden death while others lose their entanglement only after infinite time. Thus, there are two possible routes of entanglement decay, namely sudden death and asymptotic decay. We demonstrate that starting with an initial condition which leads to finite-time disentanglement, we can alter the future course of entanglement by local unitary actions. In other words, it is possible to put the quantum states on other track of decay once they are on a particular route of decay. We show that one can accelerate or delay sudden death. However, there is a critical time such that if local actions are taken before that critical time then sudden death can be delayed to infinity. Any local unitary action taken after that critical time can only accelerate or delay sudden death.

In finite-temparature reservoirs, we demonstrate that a whole class of entangled states exhibit sudden death. This conclusion is valid if at least one of the reservoirs is at finite-temperature. However, we show that we can still hasten or delay sudden death by local unitary transformations up to some finite time.

We also study sudden death for qubit-qutrit systems. Similar to qubit-qubit systems, some states exhibit sudden death while others do not. However, the process of disentanglement can be effected due to existence of quantum interference between excited levels of qutrit. We show that it is possible to hasten, delay, or avoid sudden death by local unitary actions taken later in time.

Zusammenfassung

Diese Arbeit ist eine theoretische Untersuchung der Verschränkungsdynamik und ihrer Steuerung für Qubit-Qubit- und Qubit-Qutrit-Systeme. Insbesondere haben wir unseren Blick auf den Zerfall der Verschränkung in Quantensystemen gerichtet, wenn sie mit dissipativen Umgebungen wechselwirken. Qubit-Qubit-Verschränkung kann bei einer Wechselwirkung mit statistisch unabhängigen Vakuumreservoirs plötzlich verschwinden. Diese Aufhebung der Verschränkung in endlicher Zeit wird *plötzlicher Verschränkungstod* genannt. Wir haben den plötzlichen Verschränkungstod für Qubit-Qubit- und Qubit-Qutrit-Systeme untersucht, die mit statistisch unabhängigen Reservoirs am absoluten Nullpunkt und bei endlicher Tematur wechselwirken. Wir haben festgestellt, daß für Reservoirs am absoluten Nullpunkt einige Quantenzustände den plötzlichen Verschränkungstod erleiden, während andere ihre Verschränkung erst nach unendlicher Zeit verlieren. Dies bedeutet, daß es zwei mögliche Wege für den Zerfall der Verschränkung gibt, d.h. der plötzliche Verschränkungstod und der asymptotische Zerfall. Wir haben gezeigt, dass wir den zukünftigen Weg der Verschränkung mittels lokal-unitärer Operationen verändern können, auch wenn die Anfangsbedingungen zu einem Aufheben der Verschränkung in endlicher Zeit führen würden. Es ist mit anderen Worten möglich, die Quantenzustände auf einen anderen Weg zu schicken, wenn sie sich bereits auf einem bestimmten Zerfallsweg befinden. Interessanterweise können wir den plötzlichen Verschränkungstod beschleunigen oder verzögern. Es gibt jedoch einen kritischen Zeitpunkt derart, daß, wenn die lokal-unitäre Operationen vor diesem Zeitpunkt angewendet werden, der Verschränkungstod bis ins Unendliche hinausgezögert werden kann. Jede lokal-unitäre Operation nach diesem kritischen Zeitpunkt kann den plötzlichen Verschränkungstod nur beschleunigen oder verzögern.

Für Reservoirs mit endlicher Temperatur haben wir festgestellt, daß alle X-Zustände den plötzlichen Verschränkungstod erleiden. Diese Ergebnis ist gültig, wenn mindestens eines der Reservoirs eine endliche Temperatur besitzt. Wir haben jedoch gezeigt, daß wir den plötzlichen Verschränkungstod immer noch bis zu einer endlichen Zeit beschleunigen oder hinauszögern können. Wir haben den plötzlichen Verschränkungstod auch für Qubit-Qutrit-Systeme untersucht. Ähnlich wie bei Qubit-Qubit-Systemen erleiden einige Zustände den plötzlichen Verschränkungstod. Der Verlauf des Zerfalls der Verschränkung kann durch das Vorliegen von Quanteninterferenz zwischen den angeregten Zuständen des Qutrits erfolgen. Wir haben gezeigt, daß es möglich ist, den plötzlichen Verschränkungstod durch lokal-unitäre Operationen zu einem späteren Zeitpunkt zu beschleunigen, zu verzögern oder vollständig zu vermeiden.

Contents

1 Introduction   1

2 Entanglement: From philosophy to technology   5
    2.1 Entangled and separable quantum states 6
    2.2 Measures of entanglement 8

3 Dynamics of open quantum systems   13
    3.1 Dynamics of a quantum system 14
       3.1.1 The Liouville-Von Neumann equation 14
       3.1.2 Interaction picture 15
       3.1.3 Dynamics of open systems 16
    3.2 Quantum Markov processes 18
       3.2.1 Quantum dynamical semigroups 18
       3.2.2 The Markovian quantum master equation 20
       3.2.3 Born and Markov approximations 23
    3.3 The quantum optical master equation 24
       3.3.1 Matter-field interaction Hamiltonian 24
       3.3.2 Atomic decay by thermal reservoirs 28

4 Entanglement sudden death   31
    4.1 Sudden death via amplitude damping 32
    4.2 Sudden death via phase damping 34
       4.2.1 Disentanglement due to global collective noise 34
       4.2.2 Disentanglement due to local noise 37
    4.3 Further recent investigations 38

5 Hastening, delaying or avoiding entanglement sudden death of qubit-qubit systems   43
    5.1 Numerical evidence for hastening, delaying or avoiding sudden death 44

	5.2	Manipulating entanglement sudden death in zero- and finite-temperature reservoirs 50
		5.2.1 Open-system dynamics of two-qubits coupled to statistically independent thermal reservoirs 50
		5.2.2 The Peres-Horodecki criterion and entanglement sudden death 53
		5.2.3 Two-qubit X-states and quantum control of entanglement sudden death 58
	5.3	Delaying, hastening, and avoiding sudden death of entanglement in statistically independent vacuum reservoirs 59
	5.4	Hastening and delaying sudden death in statistically independent thermal reservoirs 62

6 Manipulating entanglement sudden death of qubit-qutrit systems   67

6.1 Entanglement sudden death of qubit-qutrit systems by amplitude damping 67
    6.1.1 Maximally entangled pure states for $2 \otimes 3$ systems . . . 68
    6.1.2 Three-level atom and quantum interference 69
    6.1.3 Physical Model 72
    6.1.4 Dynamical process of disentanglement 73
6.2 Delaying, hastening and avoiding sudden death 78
    6.2.1 Effect of local unitary operations on entanglement dynamics 79
    6.2.2 Werner-like states 81
    6.2.3 Avoiding finite-time disentanglement during the interaction process 86
6.3 Asymptotic states and interference 88
6.4 Sudden death of qubit-qutrit systems by phase damping . . . 89

7 Summary and conclusion   95

Bibliography   97

Acknowledgments   105

Chapter 1

Introduction

Quantum physics is an accurate description of Nature. The predictions of quantum mechanics have been realized in numerous experiments. Despite its growing success, quantum mechanics offer certain intriguing and counter-intuitive features, e.g. quantum interference and quantum entanglement. These two fundamental notions have no classical analog and are at the heart of quantum mechanics. In early 1980s, it was discovered that it is not possible to clone an unknown quantum state (*no-cloning theorem*). This result is one of the earliest results of recently emerging field of quantum information and quantum computation. This field promises new technologies like quantum cryptography, quantum teleportation, quantum dense coding, and quantum computation. All these effect are not possible in classical physics. Quantum computation and quantum information is mainly based on the ability to have control over single quantum systems. For example, many techniques have been developed for trapping a single atom (ion) in a trap, and probing its different aspects with precision. After having control over single quantum systems, the task of information transmission and processing can be accomplished.

Many applications of quantum information rely on quantum entanglement. Entanglement is one of the surprising features of quantum mechanics, which gives us a description for multipartite quantum systems whereas such description does not exist for each individual system alone [1, 2]. Entanglement has turned out to be a precious resource for quantum technology. The type of correlation associated with entanglement is qualitatively different from any other known correlations. Entanglement may be shared among pairs of atoms, photons, etc., even though they may be remotely located and do not interact with each other. However, all quantum systems interact with their respective surroundings. Such unavoidable interactions cause the decay of coherence. Such decay has been recognized as *decoherence*. Decoherence may result in the degradation of quantum entanglement shared by two or more parties. It is important to study and understand the dynamics of entanglement under the influence of dissipative environments for realistic quantum information processing. Ideally, we demand that entanglement should be maintained for sufficiently long times to allow designed tasks of quantum information processing.

The main phenomenon investigated in this thesis is a special type of decoherence. It is well known that decoherence gradually eliminates quantum coherence of single quantum systems such as a spin, or an atom. The coherence of multipartite quantum systems is called *global* coherence and it is related to quantum entanglement. Decoherence leads to loss of entanglement and consequently entanglement-dependent applications of quantum information may not be realized experimentally. It has been observed that two-qubits entanglement may be lost in a very different way compared to local decoherence measured by the decay of off-diagonal elements of the density matrix of either qubit. Yu and Eberly have reported the surprising observation that the presence of either pure vacuum noise or even classical noise can cause entanglement to decay to zero in finite time although local coherences decay in infinite time [61, 63]. This effect is called "*entanglement sudden death*" (ESD), or finite-time disentanglement, or early-stage disentanglement. Such dissipation is a special form of decay which attacks only quantum entanglement as it has not been previously encountered in the dissipation of other physical correlations [130]. Entanglement sudden death has been predicted in numerous theoretical studies in a wide variety of cases, such as atomic qubits [83], photonic and spin qubits [89], continuous Gaussian states [58, 59], finite spin chains [81], multipartite systems [117], etc. This effect has been detected in laboratory in two optical setups [64, 65] and in an atomic ensemble [66], confirming its experimental relevance. Despite numerous theoretical studies and experimental observations, we still lack a deep understanding of sudden death dynamics.

Similar studies for qubit-qutrit systems, qutrit-qutrit systems and some special states of qudit-qudit systems indicate that sudden death is a generic phenomenon. We need to think of some measures to protect quantum information processing from this possible threat. In this regard it is important to understand the behavior of decoherence and the dynamics of entanglement in various physical situations. In addition, it is desired to have a control on dynamics of entanglement for quantum information processing.

Clearly, sudden death of entanglement can seriously affect various applications of quantum information processing. Therefore, it would be of interest if we could take suitable actions when faced with the prospect of loss of entanglement to postpone that end. We restrict ourselves here to finite dimensions and bipartite quantum states. More specifically, we consider qubit-qubit and qubit-qutrit systems to study such a possibility. Some studies on changing the initial state into an equivalent but more robust entangled state have been carried out. However, we deal with the more direct question that for a given initial state and a setup which will disentangle in a finite time, can we take suitable actions later to change the future dynamics of entanglement? Indeed we can do that. We show that simple local unitary operations can alter the time of disentanglement. We demonstrate that for certain two-qubit entangled states namely X-states (properties of X-states are described in Chapter 5) interacting with statistically independent vacuum reservoirs, simple local unitary operations can completely avoid sudden death of entanglement. However, there is always some critical time for taking such local actions and if local actions are taken before that critical time then sudden death can be completely averted. For local actions taken after this critical time, two interesting possibilities exist i.e., either sudden death is delayed up to some finite time or it is accelerated.

We show that all X-states interacting with statistically independent thermal reservoirs exhibit finite-time disentanglement. In this case there does not exist any local unitary operation which can completely avoid sudden death. However, depending upon the time of applying local unitary operations, sudden death can be accelerated or delayed only up to some finite time. Such manipulation of the time of sudden death depends on the amount of temperature in reservoirs. If we lower the temperature then sudden death can be delayed to longer times and vice versa.

We study entanglement sudden death of qubit-qutrit systems as well. We found that similar to qubit-qubit systems, some states exhibit sudden death while others do not. We show that it is always possible to manipulate sudden death via local unitary actions.

The outlines of this thesis are as follows: In Chapter 2, we discuss the history and importance of entanglement for quantum computation and quantum information. We describe the separability (entanglement) problem in a simple way. We mention some measures of entanglement for bipartite states. In Chapter 3, we build the mathematical machinery to study the dynamics of open systems, i.e., we describe the theory of open systems and derive the general form of the *master equation*. We also discuss various approximations used in this thesis and derive the *quantum optical master equation*. The approximations bring much simplicity to the master equation and make it possible to handle analytically some bipartite quantum systems. Chapter 4 deals with the introduction of entanglement sudden death in two particular cases, i.e., via amplitude damping and phase damping. In

Chapter 5, we analyze sudden death of qubit-qubit systems interacting with statistically independent reservoirs at zero- and finite-temperature. We also discuss that we can hasten, delay, or avoid sudden death if we apply suitably chosen unitary transformations to both subsystems. In Chapter 6, we study the similar analysis as in Chapter 5 for qubit-qutrit systems. We conclude our thesis in Chapter 7 and provide references at the end.

Chapter 2

Entanglement: From philosophy to technology

Entanglement is one of the surprising and counter-intuitive phenomena of quantum mechanics. Schrödinger coined the term "Verschränkung" [1] for this nonclassical feature of multipartite physical systems. Entanglement is a purely quantum mechanical phenomenon and has no analog in classical physics. Historically, Einstein *et al.* [2] questioned the legitimacy of quantum mechanics due to entangled states. They could not apprehend this peculiar trait of quantum states and concluded that quantum mechanics is not a complete physical theory. Bohr criticized their arguments by presenting a different interpretation of *locality* and *reality* and stressed the completeness of quantum theory [3]. Entanglement was considered as a fancy mathematical entity, which could only be a subject of discussion between philosophers. In 1964, Bell succeeded to show that the statistical predictions of quantum mechanics, for certain spatially separated but correlated two-particle systems, are incompatible with a large class of deterministic local theories [4]. Bell was able to construct a mathematical relation for all correlations that can exist between the two outcomes of two distant systems, which satisfy the assumptions of *locality* and *reality*. Certain entangled states violate this mathematical relation and hence establish the non-local nature of quantum states. Bell's theorem (also called Bell's inequality) was later extended by Clauser *et al.* [5] in a form more suitable for providing an experimental test for all local hidden-variable theories. With the advancement of technology, soon it was possible to test this idea in laboratory. The world was surprised by the experimental results in favor of quantum mechanics. The pioneer experimental results testing Bell's inequalities were in excellent agreement with the predictions of quantum mechanics [6, 7, 8, 9, 10]. The experimental evidences with improved techniques (hence closing nearly all loopholes) continue to support quantum mechanics up to this day. More recently, the violation of local realism with freedom of choices has been shown to support quantum mechanics [11].

In the last two decades of the 20th century the philosophical discussion on entanglement turned into its technological aspects. In 1984, Bennett and Brassard introduced the interesting field of quantum cryptography [12]. Deutsch and others came up with the idea of quantum computation [13, 14, 15, 16]. Moreover, quantum cryptography based on Bell's theorem [17], quantum dense coding [18], and quantum teleportation [19] were also predicted. All these quantum effects are based on entangled states of two qubits. All these effects have been demonstrated in laboratory [20, 21, 22, 23, 24, 25, 26, 27].

All of the above mentioned discoveries supported with numerous experimental evidences lead to a new interdisciplinary area of research called *quantum information* [28, 29, 30, 31, 32, 33]. Quantum information deals with entanglement as a central resource. The theory of entanglement generally deals with central problems like: i) detection of entanglement both in theory and in laboratory; ii) characterization, control and quantification of entanglement; iii) addressing the unavoidable process of disentanglement [34]. In this thesis we investigate the degradation of entanglement interacting with independent dissipative environments.

2.1 Entangled and separable quantum states

A fundamental question in quantum information may be the identification of correlations existing between different quantum systems. How can one say with certainty that a given multipartite quantum state contains entanglement? The answer to this question is non-trivial. Even for the simpler case of bipartite systems, classification of quantum states into separable and entangled states is not easy. To determine separability (entanglement) of a given quantum state is itself an area of research which has been extensively explored, see Refs. [34, 35] and references therein. We will provide a simple definition of entanglement and restrict ourselves to bipartite quantum systems, in particular qubit-qubit and qubit-qutrit systems which are relevant for our work.

The simplest definition of separability (entanglement) is for pure bipartite quantum states. Let $\mathcal{H}$ be a Hilbert space such that $\mathcal{H} = \mathcal{H}_A \otimes \mathcal{H}_B \cong \mathbb{C}^{d_1} \otimes \mathbb{C}^{d_2}$ (with integers d_1, $d_2 \geq 2$). Any bipartite pure state $|\Psi_{AB}\rangle \in \mathcal{H}$ is called separable (entangled) if and only if it can be (cannot be) written as a direct product of two vectors corresponding to the Hilbert spaces of the

subsystems, i.e.,
$$|\Psi_{AB}\rangle = |\psi_A\rangle \otimes |\phi_B\rangle, \qquad (2.1)$$
where $|\psi_A\rangle \in \mathcal{H}_A$, and $|\phi_B\rangle \in \mathcal{H}_B$.

Another simple way to determine the separability of pure states is based on the *Schmidt decomposition*. We only provide the main theorem as the proof can be found in any standard text on quantum information [29].

Schmidt decomposition 2.1.1 *Let $|\Psi_{AB}\rangle \in \mathcal{H}$ be a bipartite pure state, then there exist orthonormal states $|e_i\rangle \in \mathcal{H}_A$ and $|f_i\rangle \in \mathcal{H}_B$ such that*
$$|\Psi_{AB}\rangle = \sum_i \lambda_i |e_i\rangle \otimes |f_i\rangle, \qquad (2.2)$$
with $\lambda_i \geq 0$ and $\sum_i |\lambda_i|^2 = 1$. The coefficients λ_i are the Schmidt coefficients. The number of nonzero Schmidt coefficients is referred to as Schmidt rank of $|\Psi_{AB}\rangle$. The state $|\Psi_{AB}\rangle$ is separable if and only if it has Schmidt rank one.

Due to decoherence, we usually deal with mixed states rather than pure states. For mixed states, the characterization of separability is not so easy. However, it is defined that any bipartite mixed state ρ_{AB} defined on $\mathcal{H} = \mathcal{H}_A \otimes \mathcal{H}_B$ is separable [36] if and only if it can be written as
$$\rho_{AB} = \sum_{i=1}^{n} p_i \rho_A^i \otimes \rho_B^i, \qquad (2.3)$$
where $p_i \geq 0$ and $\sum_i p_i = 1$, $\rho_A^i \in \mathcal{H}_A$ and $\rho_B^i \in \mathcal{H}_B$. For a given mixed state ρ_{AB}, it is very hard to check its separability (entanglement) directly. It is quite difficult to determine separability of a given mixed state and simple criteria exist only in some special cases. In this thesis, we are dealing with quantum states defined in the Hilbert spaces of dimensions 4 and 6, namely qubit-qubit ($2 \otimes 2$) and qubit-qutrit ($2 \otimes 3$) systems, respectively. For these dimensions of the Hilbert spaces, there exists an operational criterion, which is both necessary and sufficient to check separability (entanglement) of quantum states. This criterion provided by Peres [37] is called the positive partial transpose (PPT) criterion. It states that *if a quantum state ρ_{AB} is separable then the matrix ρ_{AB}^{PT}, obtained after taking the partial transpose of ρ_{AB}, is also a valid quantum state*. It was shown by Horodecki *et al* [38] that for qubit-qubit and qubit-qutrit systems, the Peres criterion is both necessary and sufficient. This criterion is often called the Peres-Horodecki criterion for separability. The partial transpose means that we take the transpose with respect to indices of any one of the subsystems A or B.

For some fixed orthonormal product basis, the matrix elements of $\rho_{AB}^{T_B}$ are defined by:

$$\langle m|\langle \mu|\rho_{AB}^{T_B}|n\rangle|\nu\rangle \equiv \langle m|\langle \nu|\rho_{AB}|n\rangle|\mu\rangle, \qquad (2.4)$$

where the operation T_B means transposition of indices corresponding to the subsystem B.

The Peres-Horodecki criterion was also shown to be necessary and sufficient for low rank states [39], pure states [40], rank two states [41], and rank three states [39]. However, for Hilbert spaces of dimension (≥ 8), there are some entangled states having positive partial transpose [35, 42]. Such peculiar entangled states are called *bound entangled states* (BES), because their entanglement cannot be distilled to pure entangled states. These observations imply that the set of PPT states contain both separable and entangled states. However, it is certain that if a quantum state has negative partial transpose (NPT) then the state is entangled. NPT means that the matrix after taking partial transpose must have at least one negative eigenvalue. There is a conjecture (on the basis of numerical evidence) for the existence of bound entangled states having negative partial transpose [43, 44]. However, the conclusive analytical evidence is still missing.

As mentioned earlier, our main discussion in this thesis will focus on qubit-qubit and qubit-qutrit systems and the Peres-Horodecki criterion guarantees that for these systems all PPT states are separable. After recognition of all entangled (separable) states for our systems of interest, we can now move to quantify entanglement.

2.2 Measures of entanglement

To quantify the amount of entanglement of a given quantum state is one of the central and important issues of quantum information. Much effort has been devoted to this area of research and several useful measures of entanglement have been worked out for bipartite and multipartite quantum systems. We will restrict our discussion only to bipartite quantum systems by providing some references for multipartite systems. There exist several proposed measures of entanglement. However, this discussion is not the main theme of this thesis therefore we briefly discuss some measures of entanglement.

A general measure of entanglement has to be an *entanglement monotone* (E). An entanglement monotone is a positive functional that maps entangled states to positive real numbers. For separable states, an entanglement monotone must be zero and it must have maximum value for maximally entangled states. Any entanglement monotone should satisfy five physically motivated properties (see Ref. [45] and references therein). Let $\mathcal{S}(\mathcal{H})$ be the set of all quantum states on the Hilbert space $\mathcal{H}$, and $\mathcal{D}(\mathcal{H})$ be the set of separable states, then the properties for an entanglement monotone are:

(i) $E : \mathcal{S}(\mathcal{H}) \to \mathbb{R}$ is a positive functional, and $E(\sigma) = 0$ for any separable state $\sigma \in \mathcal{D}(\mathcal{H})$.

(ii) E is a convex functional, i.e.,

$$E\Big(\sum_{i=1}^{n} p_i \sigma_i\Big) \leq \sum_{i=1}^{n} p_i E(\sigma_i), \qquad (2.5)$$

for $p_i \in [0,1]$ and $\sigma_i \in \mathcal{S}(\mathcal{H})$, $i = 1, \ldots, n$, with $\sum_{i=1}^{n} p_i = 1$.

(iii) E is monotone under local operations and classical communication (LOCC): This means if

$$\sigma_i = \frac{\sum_j (A_{i,j} \otimes I_B) \sigma (A_{i,j} \otimes I_B)^\dagger}{p_i}, \qquad i = 1, \ldots, k, \qquad (2.6)$$

with probability $p_i = \mathrm{tr}\{\sum_j A_{i,j} \sigma A_{i,j}^\dagger\}$, and $\sum_{i,j}^{k} A_{i,j}^\dagger A_{i,j} = I_A$, then

$$E(\sigma) \geq \sum_{i=1}^{k} p_i E(\sigma_i). \qquad (2.7)$$

Eq. (2.7) implies that the degree of entanglement does not increase under LOCC. Property (iii) also leads to an invariance under local unitary transformations, i.e., $E(U\rho U^\dagger) = E(\rho)$ for all $\rho \in \mathcal{S}(\mathcal{H})$ and all local unitary operations $U = U_A \otimes U_B : \mathcal{H} \to \mathcal{H}$.

(iv) E is weakly additive, i.e., $E\big(|\psi\rangle\langle\psi|^{\otimes n}\big) = nE(|\psi\rangle\langle\psi|)$ for all $|\psi\rangle \in \mathcal{H}$ and all $n \in \mathbb{N}$.

(v) E is weakly continuous, i.e., for a given $|\psi\rangle \in \mathcal{H}$, let (σ_n) be a series of states $\sigma_n \in \mathcal{S}(\mathcal{H}^{\otimes n})$ with the property that $\lim_{n\to\infty} \| |\psi\rangle\langle\psi|^{\otimes n} - \sigma_n \| = 0$, where $\|.\|$ is the trace norm[1], then E satisfies

$$\lim_{n\to\infty} \frac{1}{n} | E(|\psi\rangle\langle\psi|^{\otimes n}) - E(\sigma_n) | = 0. \qquad (2.8)$$

Next we describe some proposed measures of entanglement.

---

[1] The trace norm of a matrix is defined as $\|A\| = \mathrm{tr}|A| = \mathrm{tr}[\sqrt{A^\dagger A}]$.

Von Neumann entropy

Von Neumann entropy of the reduced quantum state $\rho_B = \mathrm{tr}_A(|\psi\rangle\langle\psi|)$ is the uniquely defined entanglement measure for pure states of bipartite quantum systems [29]. It is defined by

$$E(|\psi\rangle\langle\psi|) = S(\mathrm{tr}_A\{|\psi\rangle\langle\psi|\}) = S(\mathrm{tr}_B\{|\psi\rangle\langle\psi|\}), \qquad (2.9)$$

where $\mathrm{tr}_{A(B)}$ is partial trace over indices of system $A(B)$ and S is the *Von Neumann entropy*.

Distillable entanglement

Distillable entanglement is defined as the maximal number of maximally entangled states that can be extracted from many copies of a given entangled state σ by means of local operations and classical communication (LOCC). We can transform a certain number of non-maximally entangled states into a smaller number of approximately maximally entangled states with the use of LOCC [46, 47]. Such an extraction is similar to "distilling". Let $D_{\leftrightarrow}$ denote *distillable entanglement* [48, 49] with respect to LOCC, also called *two-way distillable entanglement*.

For pure states $S(\mathrm{tr}_A\{|\psi\rangle\langle\psi|\})$ quantifies the amount of EPR pairs contained asymptotically in the state $|\psi\rangle\langle\psi|$, i. e.,

$$D_{\leftrightarrow} = S(\mathrm{tr}_A\{|\psi\rangle\langle\psi|\}) = S(\mathrm{tr}_B\{|\psi\rangle\langle\psi|\}). \qquad (2.10)$$

For a general mixed state, it is hard to evaluate this measure [48, 49]. For bound entangled states, $D_{\leftrightarrow} = 0$.

Entanglement of formation

Entanglement of formation is defined as the number of maximally entangled states required to prepare copies of a particular state in the asymptotic limit of many copies [50]. For pure states, it is given by

$$E_F(|\psi\rangle\langle\psi|) = S(\mathrm{tr}_A\{|\psi\rangle\langle\psi|\}). \qquad (2.11)$$

This definition can be extended to mixed states by

$$E_F(\sigma) = \min \sum_i \mu_i E(|\psi_i\rangle\langle\psi_i|), \qquad (2.12)$$

where the minimum is taken over all possible decompositions

$$\sigma = \sum_i \mu_i |\psi_i\rangle\langle\psi_i|. \qquad (2.13)$$

It is hard to evaluate E_F for general mixed states due to the complicated minimization procedure. However, for quantum states of two qubits, a general formula has been worked out to compute *entanglement of formation* [51, 52].

Negativity

Negativity is easy to compute and it does not involve a complicated minimization over a high dimensional space. It was first investigated by Życzkowski *et al.* [53]. It is connected with the Peres-Horodecki criterion and is defined by

$$N(\sigma) = \|\sigma^{T_B}\| - 1. \tag{2.14}$$

Thus it is twice the sum of the absolute values of all the negative eigenvalues of the partially transposed density matrix [54]. *Negativity* varies from $N = 0$ for the PPT states (hence separable for qubit-qubit and qubit-qutrit systems) to $N = 1$ for maximally entangled states, e.g. Bell states. Therefore, *negativity* is a reasonable entanglement measure for qubit-qubit and qubit-qutrit systems.

Concurrence

Concurrence was initially defined for $2 \otimes 2$ systems [52], however some generalizations do exist for higher dimensions of the Hilbert spaces [55]. Let ρ be a density matrix of a $2 \otimes 2$ system and let $\lambda_1, \lambda_2, \lambda_3, \lambda_4$ be the eigenvalues of the matrix

$$\zeta \equiv \rho(\sigma_y^A \otimes \sigma_y^B)\rho^*(\sigma_y^A \otimes \sigma_y^B), \tag{2.15}$$

arranged in decreasing order. Here ρ^* denotes the complex conjugate of ρ in the corresponding basis and σ_y is the standard Pauli matrix. *Concurrence* is then given by

$$C(\rho) = \max\left\{0, \sqrt{\lambda_1} - \sqrt{\lambda_2} - \sqrt{\lambda_3} - \sqrt{\lambda_4}\right\}. \tag{2.16}$$

Concurrence varies from $C = 0$ for a separable state to $C = 1$ for a maximally entangled state, e.g. a Bell state.

Chapter 3

Dynamics of open quantum systems

An open system is defined as one which has interactions with an environment whose dynamics we want to average over. The system of interest is called the *principal* system while any other system is called the *environment*. Those quantum systems which do not suffer any unwanted interactions with an environment are called *closed* systems. However, a closed system is an idealization and there does not exist any closed system in Nature except probably the universe itself. Many interesting and fascinating applications of quantum information deal with closed quantum systems where the efficiency of information processing reaches its maximum value. Examples are quantum key distribution [17] and quantum teleportation [19]. The idealistic conclusions about these quantum feats are effected by the fact that real quantum systems always suffer from unwanted interactions with their environments [29]. These unwanted interactions appear as quantum *noise*. Quantum noise can seriously effect applications of quantum information processing. To understand and control such noise processes is one of the central issue in quantum information and quantum computation [29].

The theory of open quantum systems has been discussed extensively in the literature (see Ref. [56] and references therein). Contrary to the case of a closed system, quantum dynamics of an open system does not, in general, follow unitary time evolution. The dynamics of an open system can sometimes be formulated by an appropriate differential equation of motion for its density operator. This equation is called the *quantum master equation* which may be quite useful in many cases. We will restrict ourselves to general Markovian dynamics in which the environmental excitations decay over short times and information regarding past time evolution is destroyed.

This Chapter is organized as follows. In Section 3.1, we discuss the dynamics of quantum systems. The quantum Markov processes and the Markovian quantum master equation along with the Born- and Markov-approximations are discussed in Section 3.2. The quantum optical master equation is derived in Section 3.3, where we concentrate on the limiting case of weak-coupling between radiation and matter.

3.1 Dynamics of a quantum system

3.1.1 The Liouville-Von Neumann equation

Quantum mechanics tells us that the time evolution of a state vector $|\psi(t)\rangle$ of a closed quantum system is governed by the Schrödinger equation

$$i\hbar \frac{d}{dt}|\psi(t)\rangle = H|\psi(t)\rangle, \qquad (3.1)$$

where H is the Hamiltonian of the system. The solution of Eq. (3.1) may be written by

$$|\psi(t)\rangle = \exp\left[-\frac{i}{\hbar} H(t - t_0)\right] |\psi(t_0)\rangle. \qquad (3.2)$$

For mixed states, the corresponding statistical ensemble is characterized by a density operator ρ. Let the state of the system at an initial time t_0 be given by the density operator

$$\rho(t_0) = \sum_j p_j |\psi_j(t_0)\rangle\langle\psi_j(t_0)|, \qquad (3.3)$$

where p_j are the positive weights and $|\psi_j(t_0)\rangle$ are the corresponding state vectors. The time evolution of the density operator is given by

$$\rho(t) = U(t, t_0)\, \rho(t_0)\, U^\dagger(t, t_0). \qquad (3.4)$$

The equation of motion for the density operator is given by

$$\frac{d}{dt}\rho(t) = -\frac{i}{\hbar} [H, \rho(t)]. \qquad (3.5)$$

Eq. (3.5) is often called the *Liouville* or *Von Neumann* equation of motion. The square brackets on the right hand side of Eq. (3.5) define the commutator[1] between operators H and $\rho(t)$.

---
[1] The commutator between two arbitrary operators A and B is defined as $[A, B] := AB - BA$.

In analogy to the equation of motion for probability distribution in classical statistical mechanics, the Von Neumann equation is sometime written as

$$\frac{d}{dt}\rho(t) = \mathcal{L}\,\rho(t)\,, \tag{3.6}$$

where $\mathcal{L}$ is the *Liouville operator* defined through the condition that $\mathcal{L}\rho$ is equal to $-i/\hbar$ times the commutator of H with $\rho(t)$. $\mathcal{L}$ is also called a Liouville *super-operator* because it acts on an operator to yield another operator. For a time-independent Hamiltonian the Liouville super-operator is also time-independent and we have

$$\rho(t) = \exp[\mathcal{L}(t-t_0)]\,\rho(t_0)\,. \tag{3.7}$$

3.1.2 Interaction picture

The interaction picture is a general picture and the Schrödinger picture is a limiting case of it. We can write the Hamiltonian of the system as the sum of two parts

$$H(t) = H_0 + \hat{H}_I(t)\,. \tag{3.8}$$

Here, H_0 is the time independent sum of energies of two systems in the absence of interaction. $\hat{H}_I(t)$ is the Hamiltonian describing the interaction between the systems. The expectation value of a Schrödinger observable $O(t)$ at time t is given by

$$\langle O(t) \rangle = \text{tr}\{O(t)\,U(t,t_0)\rho(t_0)U^\dagger(t,t_0)\}\,, \tag{3.9}$$

where $\rho(t_0)$ is the state of the system at time t_0.

Introducing the unitary time evolution operators

$$U_0(t,t_0) \equiv \exp\bigl[-\frac{i}{\hbar}H_0(t-t_0)\bigr]\,, \tag{3.10}$$

and

$$U_I(t,t_0) \equiv U_0^\dagger(t,t_0)U(t,t_0)\,, \tag{3.11}$$

the expectation value Eq. (3.9) can be written as

$$\begin{aligned}\langle O(t) \rangle &= \text{tr}\{U_0^\dagger(t,t_0)O(t)U_0(t,t_0)U_I(t,t_0)\rho(t_0)U_I^\dagger(t,t_0)\} \\ &\equiv \text{tr}\{O_I(t)\rho_I(t)\}\,,\end{aligned} \tag{3.12}$$

where we have introduced $O_I(t)$ as the interaction picture operator

$$O_I(t) \equiv U_0^\dagger(t,t_0)O(t)U_0(t,t_0),\quad(3.13)$$

and $\rho_I(t)$ as the interaction picture density operator

$$\rho_I(t) \equiv U_I(t,t_0)\rho(t_0)U_I^\dagger(t,t_0).\quad(3.14)$$

For the case of vanishing free Hamiltonian $H_0 = 0$, we have $H(t) = \hat{H}_I(t)$ such that $U_0(t,t_0) = I$ and $U_I(t,t_0) = U(t,t_0)$, and we obtain the Schrödinger picture.

The interaction picture time-evolution operator $U_I(t,t_0)$ is the solution of the differential equation

$$i\hbar\frac{\delta}{\delta t}U_I(t,t_0) = H_I(t)U_I(t,t_0),\quad(3.15)$$

with the initial condition $U_I(t_0,t_0) = I$. In Eq. (3.15), we have denoted the interaction Hamiltonian in the interaction picture by

$$H_I(t) \equiv U_0^\dagger(t,t_0)\hat{H}_I(t)U_0(t,t_0).\quad(3.16)$$

The corresponding Von Neumann equation in the interaction picture is given by

$$\frac{d}{dt}\rho_I(t) = -\frac{i}{\hbar}[H_I(t),\rho_I(t)].\quad(3.17)$$

The integral form of the Von Neumann equation in the interaction picture is given by

$$\rho_I(t) = \rho_I(t_0) - \frac{i}{\hbar}\int_{t_0}^t dx\,[H_I(x),\rho_I(x)].\quad(3.18)$$

3.1.3 Dynamics of open systems

An open system is a quantum system S which is coupled to another quantum system E called environment. An open system represents a subsystem of the combined system $S + E$. In most of the cases, it is assumed that the total system is closed and follows the Hamiltonian dynamics. The state of the system S changes as a consequence of its internal dynamics and due to interaction with the environment. This interaction leads to certain system-environment correlations and corresponding changes of the system S can no longer be represented in terms of unitary time evolution. The dynamics of the system S is often called as the *reduced system dynamics*.

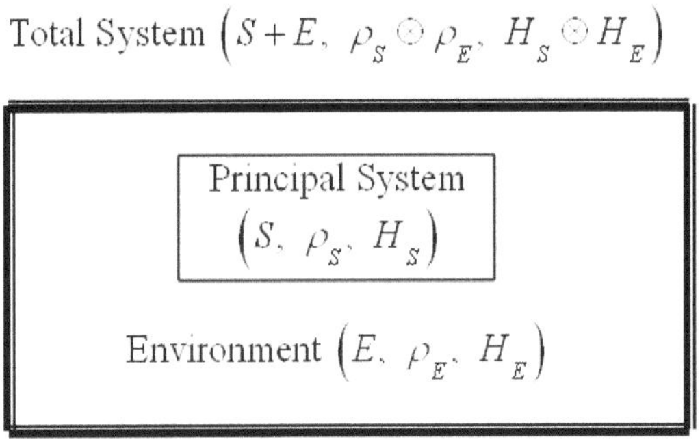

Total System $\left(S+E,\ \rho_S \otimes \rho_E,\ H_S \otimes H_E\right)$

```
┌─────────────────────────────────────────────┐
│           ┌───────────────────┐             │
│           │  Principal System │             │
│           │   (S, ρ_S, H_S)   │             │
│           └───────────────────┘             │
│                                             │
│        Environment (E, ρ_E, H_E)            │
│                                             │
└─────────────────────────────────────────────┘
```

Figure 3.1: The model of the combined system $S + E$. The principal (open) system interacts with the environment.

Let $\mathcal{H}_S$ denote the Hilbert space of the system S and $\mathcal{H}_E$ the Hilbert space of the environment E. The Hilbert space of the combined system $S+E$ is given by the tensor product space $\mathcal{H} = \mathcal{H}_S \otimes \mathcal{H}_E$. The Hamiltonian of the combined system takes the form

$$H(t) = H_S \otimes I_E + I_S \otimes H_E + \hat{H}_I(t), \tag{3.19}$$

where H_S is the Hamiltonian of the open system S, H_E is the free Hamiltonian of the environment E, and $\hat{H}_I(t)$ is the Hamiltonian describing the interaction between the system and the environment. Figure 3.1 shows the typical situation under discussion.

In many physical situations a complete mathematical model of the combined system $(S + E)$ is very complicated. The environment may be a reservoir or a heat bath consisting of infinitely many degrees of freedom and one has to solve infinitely coupled equations of motion. Even if a solution is known, one still has the problem of isolating and determining the interesting physical quantities through an average over the irrelevant degrees of freedom. More often, the modes of the environment are neither known exactly nor controllable. Therefore, a simpler description is desired in a reduced state space by applying various analytical methods and approximation techniques.

The observables of the system S are all of the form $O \otimes I_E$, where O is an operator acting on the Hilbert space $\mathcal{H}_S$ and I_E denotes the identity operator in the Hilbert space $\mathcal{H}_E$. If ρ is the state of the total system then the expectation values of all observable acting on the Hilbert space of the open system S alone are determined by

$$\langle O \rangle = \text{tr}\left\{O\,\rho_S\right\}, \tag{3.20}$$

where ρ_S is the reduced density operator of the open system S obtained by taking trace over the degrees of freedom of the environment E, i.e., $\rho_S = \text{tr}_E\{\rho\}$. The reduced density operator ρ_S is of central importance in the description of the open quantum systems.

The time-dependent reduced density operator $\rho_S(t)$ at time t is obtained from the density operator $\rho(t)$ of the total system. As the combined system evolves unitarily, we have

$$\rho_S(t) = \text{tr}_E\left\{U(t,t_0)\,\rho(t_0)\,U^\dagger(t,t_0)\right\}. \tag{3.21}$$

Similarly the equation of motion for the reduced density operator is obtained by taking trace over the environment on both sides of the Von Neumann equation of motion

$$\frac{d}{dt}\rho_S(t) = -\frac{\text{i}}{\hbar}\,\text{tr}_E\left[H(t), \rho(t)\right]. \tag{3.22}$$

3.2 Quantum Markov processes

An important property of a classical, homogeneous Markov process is the semigroup property, which is formulated in terms of a differential equation involving a time-independent generator. The extension of this idea to quantum mechanics leads to the concept of quantum dynamical semigroups and quantum Markov processes. In this section, we introduce these concepts and derive the general form of a quantum Markov master equation.

3.2.1 Quantum dynamical semigroups

The dynamics of the reduced system S (Eq. (3.22)) is in practice quite difficult to solve. However, with the condition of short environmental correlation time, we may neglect memory effects and formulate the reduced system dynamics in terms of a quantum dynamical semigroup.

First we define a dynamical map. Let us prepare the state of the total system $(S+E)$ at time $t=0$ in an uncorrelated state $\rho(0) = \rho_S(0) \otimes \rho_E$. Here, $\rho_S(0)$ is the initial state of the open system and ρ_E is the reference state of the environment. The transformation changing the reduced system from time $t=0$ to some later time $t>0$ may be written in the form

$$\rho_S(0) \mapsto \rho_S(t) = T(t)\rho_S(0) \equiv \text{tr}_E\left\{U(t,0)[\rho_S(0) \otimes \rho_E]U^\dagger(t,0)\right\}. \tag{3.23}$$

Considering the state ρ_E and the final time t to be fixed, Eq. (3.23) defines a map from the space $\mathcal{S}(\mathcal{H}_S)$ of density matrices of the reduced system into itself,

$$T(t) : \mathcal{S}(\mathcal{H}_S) \to \mathcal{S}(\mathcal{H}_S) . \tag{3.24}$$

This map, describing the state change of an open system over time t, is called a *dynamical map*. A dynamical map can be characterized completely in terms of operators acting on the Hilbert space $\mathcal{H}_S$ of the open system S. We use the spectral decomposition of the environment density operator by

$$\rho_E = \sum_i \lambda_i |\psi_i\rangle\langle\psi_i| , \tag{3.25}$$

where $|\psi_i\rangle$ form an orthonormal basis in $\mathcal{H}_E$ and λ_i are non-negative real numbers satisfying $\sum_i \lambda_i = 1$.

Eq. (3.23) yields the following representation

$$T(t)\rho_S = \sum_{i,j} K_{ij}(t)\rho_S K_{ij}^\dagger(t) , \tag{3.26}$$

where the operators K_{ij} in $\mathcal{H}_S$ are defined by

$$K_{ij}(t) = \sqrt{\lambda_j} \langle\psi_i|U(t,0)|\psi_j\rangle . \tag{3.27}$$

The dynamical map $T(t)$ is of the form of an operation describing a generalized quantum measurement and satisfies the condition

$$\sum_{i,j} K_{ij}^\dagger(t) K_{ij}(t) = I_S . \tag{3.28}$$

Based on this observation, we deduce that

$$\mathrm{tr}_S \{T(t)\rho_S\} = \mathrm{tr}_S \{\rho_S\} = 1 . \tag{3.29}$$

Therefore, a dynamical map represents a convex-linear, completely positive and trace preserving quantum operation.

Eq. (3.26) defines a dynamical map for a fixed time $t \geq 0$. However, a complete one-parameter family of maps can be constructed by allowing the time t to vary. This family of maps with $T(0) = I$ describes the whole future time evolution of the open system. If the characteristic time scales over which the reservoir correlation functions decay are much smaller than the characteristic time scale of the system evolution, it is justified to neglect memory effects in the reduced system dynamics. Similar to classical theory, we expect the Markovian-type behavior. The Markovian-type dynamics may be formulated using the semigroup property

$$T(t_1)T(t_2) = T(t_1 + t_2) , \quad t_1, t_2 \geq 0 . \tag{3.30}$$

Hence a quantum dynamical semigroup is a continuous, one-parameter family of dynamical maps satisfying the semigroup property Eq. (3.30).

3.2.2 The Markovian quantum master equation

For a given quantum dynamical semigroup, a linear map $\mathcal{L}$ under certain conditions (discussed below), allows to represent the dynamical map in the form

$$T(t) = \exp(\mathcal{L}t). \tag{3.31}$$

This equation yields a first order differential equation for the reduced density operator of the open system,

$$\frac{d}{dt}\rho_S(t) = \mathcal{L}\rho_S(t), \tag{3.32}$$

called the Markovian quantum master equation. The generator $\mathcal{L}$ of the semigroup is a super-operator.

We can construct the most general form for the generator $\mathcal{L}$. Let us consider a finite dimensional Hilbert space $\mathcal{H}_S$ with dim $\mathcal{H}_S = N$. The corresponding *Liouville space* [2] is a complex space of dimension N^2 and we choose a complete basis of orthonormal operators F_m, $m = 1, 2, \ldots, N^2$, in this space such that

$$(F_m, F_n) \equiv \text{tr}\{F_m^\dagger F_n\} = \delta_{mn}. \tag{3.33}$$

Let one of the basis operators be chosen to proportional to the identity, $F_{N^2} = (1/\sqrt{N})I_S$, such that the other basis operators are traceless, i.e., $\text{tr}\{F_m\} = 0$ for $m = 1, 2, \ldots, N^2 - 1$. Applying the completeness relation to each of the operators in Eq. (3.27), we have

$$K_{ij}(t) = \sum_{m=1}^{N^2} F_m(F_m, K_{ij}(t)). \tag{3.34}$$

Eq. (3.26) in terms of these operators is given by

$$T(t)\rho_S = \sum_{m,n=1}^{N^2} c_{mn}(t) F_m \rho_S F_n^\dagger, \tag{3.35}$$

where

$$c_{mn}(t) \equiv \sum_{ij}(F_m, K_{ij}(t))(F_n, K_{ij}(t))^*. \tag{3.36}$$

---

[2]Given some Hilbert space $\mathcal{H}$ the Liouville space is the space of Hilbert-Schmidt operators, that is the space of operators $A \in \mathcal{H}$ for which $\text{tr}(A^\dagger A)$ is finite.

The coefficient matrix $c = (c_{ij})$ is easily seen to be Hermitian and positive. Utilizing Eq. (3.35) for small time δt evolution, Eq. (3.32) can be written as

$$\mathcal{L}\rho_S = \lim_{\delta t \to 0} \frac{1}{\delta t}\{T(\delta t)\rho_S - \rho_S\}$$

$$= \lim_{\delta t \to 0} \left\{ \frac{1}{N} \frac{c_{N^2 N^2}(\delta t) - N}{\delta t} \rho_S + \frac{1}{\sqrt{N}} \sum_{m=1}^{N^2-1} \left(\frac{c_{mN^2}(\delta t)}{\delta t} F_m \rho_S + \frac{c_{N^2 m}(\delta t)}{\delta t} \rho_S F_m^\dagger \right) + \sum_{m,n=1}^{N^2-1} \frac{c_{mn}(\delta t)}{\delta t} F_m \rho_S F_n^\dagger \right\}. \quad (3.37)$$

We can define the coefficients a_{mn} by

$$a_{N^2 N^2} = \lim_{\delta t \to 0} \frac{c_{N^2 N^2}(\delta t) - N}{\delta t}, \quad (3.38)$$

$$a_{mN^2} = \lim_{\delta t \to 0} \frac{c_{mN^2}(\delta t)}{\delta t}, \quad m = 1, \ldots, N^2 - 1, \quad (3.39)$$

$$a_{mn} = \lim_{\delta t \to 0} \frac{c_{mn}(\delta t)}{\delta t}, \quad m, n = 1, \ldots, N^2 - 1, \quad (3.40)$$

and introduce the quantities

$$F = \frac{1}{\sqrt{N}} \sum_{m=1}^{N^2-1} a_{mN^2} F_m, \quad (3.41)$$

and

$$G = \frac{1}{2N} a_{N^2 N^2} I_S + \frac{1}{2}(F^\dagger + F), \quad (3.42)$$

and the Hermitian operator

$$H = \frac{\hbar}{2i}(F^\dagger - F). \quad (3.43)$$

Again the matrix formed by the coefficients a_{mn}, $m, n = 1, 2, \ldots, N^2 - 1$, is Hermitian and positive. With these definitions, we can write the generator as

$$\mathcal{L}\rho_S = -\frac{i}{\hbar}[H, \rho_S] + \{G, \rho_S\} + \sum_{m,n=1}^{N^2-1} a_{mn} F_m \rho_S F_n^\dagger. \quad (3.44)$$

The middle term on the right hand side of Eq. (3.44) defines the anti-commutator[3] between the operators G and ρ_S. As the semigroup is a trace

---
[3] The anti-commutator between two arbitrary operators A and B is defined as $\{A, B\} := AB + BA$.

preserving operator we have

$$0 = \mathrm{tr}_S\{\mathcal{L}\rho_S\} = \mathrm{tr}_S\left\{\left(2G + \sum_{m,n=1}^{N^2-1} a_{mn} F_n^\dagger F_m\right)\rho_S\right\}, \quad (3.45)$$

from which we deduce that

$$G = -\frac{1}{2}\sum_{m,n=1}^{N^2-1} a_{mn} F_n^\dagger F_m. \quad (3.46)$$

The *standard form* of the generator (3.44) is given by

$$\mathcal{L}\rho_S = -\frac{i}{\hbar}[H,\rho_S] + \sum_{m,n=1}^{N^2-1} a_{mn}\left(F_m\rho_S F_n^\dagger - \frac{1}{2}\{F_n^\dagger F_m, \rho_S\}\right). \quad (3.47)$$

Since the coefficient matrix $a = (a_{mn})$ is positive, it may be diagonalized by an appropriate unitary transformation u,

$$uau^\dagger = \begin{pmatrix} \gamma_1 & 0 & \cdots & 0 \\ 0 & \gamma_2 & \cdots & 0 \\ 0 & 0 & \ddots & 0 \\ 0 & 0 & \cdots & \gamma_{N^2-1} \end{pmatrix}, \quad (3.48)$$

where the eigenvalues γ_m are non-negative. We introduce a new set of operators A_k by

$$F_m = \sum_{k=1}^{N^2-1} u_{km} A_k, \quad (3.49)$$

and the diagonal form of the generator is obtained as

$$\mathcal{L}\rho_S = -\frac{i}{\hbar}[H,\rho_S] + \sum_{k=1}^{N^2-1} \gamma_k\left(A_k\rho_S A_k^\dagger - \frac{1}{2}A_k^\dagger A_k \rho_S - \frac{1}{2}\rho_S A_k^\dagger A_k\right). \quad (3.50)$$

This is the most general form for the generator of a quantum dynamical semigroup. The first term represents the unitary part of the dynamics generated by the Hamiltonian H. The operators A_k are usually referred as Lindblad operators. The non-negative quantities γ_k have the dimension of an inverse time provided the A_k are taken dimensionless. We will discuss later that γ_k are given in terms of certain environment correlation functions and play the role of relaxation rates for different decay modes of the open system.

It is convenient sometimes to introduce the *dissipator*

$$\mathcal{D}(\rho_S) \equiv \sum_k \gamma_k \left(A_k \rho_S A_k^\dagger - \frac{1}{2} A_k^\dagger A_k \rho_S - \frac{1}{2} \rho_S A_k^\dagger A_k \right), \tag{3.51}$$

and write the quantum master equation (3.32) in the form

$$\frac{d}{dt}\rho_S(t) = -\frac{i}{\hbar}[H, \rho_S(t)] + \mathcal{D}(\rho_S(t)). \tag{3.52}$$

3.2.3 Born and Markov approximations

The generator of a quantum dynamical semigroup is desired to derive from the Hamiltonian of the total system. This derivation can be achieved under certain assumptions discussed in this section. Let us consider a quantum system S weakly coupled to an environment E. The Hamiltonian of the total system is given by

$$H = H_S + H_E + H_I, \tag{3.53}$$

where H_S (H_E) denotes the free Hamiltonian of the system (environment) and H_I is the Hamiltonian responsible for the interaction between the system and the environment. In the interaction picture, the Von Neumann equation for the total density operator $\rho(t)$ is given by

$$\frac{d}{dt}\rho(t) = -\frac{i}{\hbar}[H_I(t), \rho(t)], \tag{3.54}$$

and its integral form is

$$\rho(t) = \rho(0) - \frac{i}{\hbar} \int_0^t dx\, [H_I(x), \rho(x)]. \tag{3.55}$$

Substituting Eq. (3.55) back into Eq. (3.54), we find the equation of motion

$$\frac{d}{dt}\rho(t) = -\frac{i}{\hbar}[H_I(t), \rho(0)] - \frac{1}{\hbar^2} \int_0^t dx\, [H_I(t), [H_I(x), \rho(x)]]. \tag{3.56}$$

If the interaction energy $H_I(t)$ is zero, the system and the environment are independent and the density operator ρ would factor as a direct product $\rho(t) = \rho_S(t) \otimes \rho_E(0)$, where the environment is assumed to be at equilibrium. As the interaction is weak, we look for a solution of the form [57]

$$\rho(t) = \rho_S(t) \otimes \rho_E(0) + \rho_c(t), \tag{3.57}$$

where $\rho_c(t)$ is of higher order in $H_I(t)$. As the reduced density operator for the system ρ_S is obtained by taking a trace over the environment coordinates, therefore $\text{tr}\{\rho_c(t)\} = 0$. This approximation is called the *Born approximation*, which assumes that the coupling between the system and the environment is weak, such that the influence of the system on the environment is small. Therefore, the density operator ρ_E of the environment is negligibly affected by the interaction and the state of the total system at time t may be approximately described by a tensor product. Inserting Eq. (3.57) into Eq. (3.56) and taking the trace over the environment coordinates and retaining terms up to order $H_I^2(t)$, we obtain

$$\frac{d}{dt}\rho_S(t) = -\frac{i}{\hbar}\text{tr}_E[H_I(t), \rho_S(0) \otimes \rho_E(0)]$$
$$-\frac{1}{\hbar^2}\text{tr}_E \int_0^t dx \Big[H_I(t), [H_I(x), \rho_S(x) \otimes \rho_E(0)]\Big]. \quad (3.58)$$

The Born approximation does not imply that there are no excitations in the environment. The *Markov approximation* provides a description on a coarse-grained time scale and the assumption that environmental excitations decay over short times which can not be resolved. In the Markov approximation, $\rho_S(x)$ is replaced by $\rho_S(t)$. This is a reasonable assumption since damping destroys memory of the past. We can write Eq. (3.58) as

$$\frac{d}{dt}\rho_S(t) = -\frac{i}{\hbar}\text{tr}_E[H_I(t), \rho_S(0) \otimes \rho_E(0)]$$
$$-\frac{1}{\hbar^2}\text{tr}_E \int_0^t dx \Big[H_I(t), [H_I(x), \rho_S(t) \otimes \rho_E(0)]\Big]. \quad (3.59)$$

This is a valid equation for a system represented by ρ_S interacting with a reservoir represented by ρ_E.

3.3 The quantum optical master equation

Quantum dynamical semigroups and quantum Markovian master equation are easily realized in quantum optics as the physical conditions underlying the Markovian approximation are very well satisfied. We discuss below atom-field interaction as an example relevant for our work.

3.3.1 Matter-field interaction Hamiltonian

We consider a quantum system e. g. an atom interacting with a quantized radiation field. The radiation field represents an environment with infinite number of degrees of freedom and the quantum atomic system is our system of interest. The quantized electric field [57] is given by

$$\mathbf{E}(\mathbf{r},t) = \sum_k \hat{\epsilon}_k \mathscr{E}_k (a_k e^{-i\nu_k t + i\mathbf{k}\cdot\mathbf{r}} + a_k^\dagger e^{i\nu_k t - i\mathbf{k}\cdot\mathbf{r}}). \tag{3.60}$$

The electric field operator is evaluated in the dipole approximation at the position of the point atom. For the atom at the origin (taking the center of mass at origin), Eq. (3.60) reduces to

$$\mathbf{E} = \sum_k \hat{\epsilon}_k \mathscr{E}_k (a_k + a_k^\dagger), \tag{3.61}$$

where $\mathscr{E}_k = (\hbar\nu_k/2\epsilon_0 V)^{1/2}$. The interaction of a radiation field $\mathbf{E}$ with a single-electron atom can be written [57] in the *dipole approximation* by

$$H = H_A + H_F - e\mathbf{r}\cdot\mathbf{E}, \tag{3.62}$$

where H_A and H_F are the energies of the atom and the field respectively, when there is no interaction and $\mathbf{r}$ is the position vector of the electron. In the dipole approximation, the field is taken to be uniform over the whole atom.

The energy H_F is given in terms of creation $a_k^\dagger$ and destruction a_k operators by

$$H_F = \sum_k \hbar\nu_k \left(a_k^\dagger a_k + \frac{1}{2}\right), \tag{3.63}$$

here ν_k is the frequency of kth mode. The atomic energy H_A and $e\mathbf{r}$ can be written in terms of the atomic transition operators $\sigma_{ij} = |i\rangle\langle j|$. This operator takes an atom from level $|j\rangle$ to level $|i\rangle$. Let $\{|i\rangle\}$ denotes a complete set of atomic energy eigenstates, such that $\sum_i |i\rangle\langle i| = 1$ and $H_A |i\rangle = E_i |i\rangle$. It then follows that

$$H_A = \sum_i E_i |i\rangle\langle i| = \sum_i E_i \sigma_{ii}, \tag{3.64}$$

and

$$e\mathbf{r} = \sum_{i,j} e|i\rangle\langle i|\mathbf{r}|j\rangle\langle j| = \sum_{i,j} \wp_{ij}\sigma_{ij}, \tag{3.65}$$

where $\wp_{ij} = e\langle i|\mathbf{r}|j\rangle$ is the electric-dipole transition matrix element. Substituting expressions of H_A, H_F, $e\mathbf{r}$, and $\mathbf{E}$ into Eq. (3.62), we get

$$H = \sum_k \hbar\nu_k a_k^\dagger a_k + \sum_i E_i \sigma_{ii} + \hbar \sum_{i,j}\sum_k g_k^{ij} \sigma_{ij}(a_k + a_k^\dagger), \tag{3.66}$$

where
$$g_k^{ij} = -(\wp_{ij} \cdot \hat{\epsilon}_k \mathscr{E}_k)/\hbar, \tag{3.67}$$
and we have subtracted the zero-point energy. For simplicity, we have decomposed the radiation field into Fourier modes in a box of volume V, with periodic boundary conditions.

We discuss the case of a two level atom with $|a\rangle$ and $|b\rangle$ defined as the excited and the ground states, respectively. For simplicity, we assume $\wp_{ab}$ to be real, i.e., $\wp_{ab} = \wp_{ba}$ and $g_k = g_k^{ab} = g_k^{ba}$. Eq. (3.66) can be written as

$$H = \sum_k \hbar \nu_k a_k^\dagger a_k + (E_a \sigma_{aa} + E_b \sigma_{bb}) + \hbar \sum_k g_k (\sigma_{ab} + \sigma_{ba})(a_k + a_k^\dagger). \tag{3.68}$$

We can simplify this relation by rewritting the second term as

$$E_a \sigma_{aa} + E_b \sigma_{bb} = \frac{1}{2}\hbar \omega (\sigma_{aa} - \sigma_{bb}) + \frac{1}{2}(E_a + E_b), \tag{3.69}$$

where $\hbar \omega = (E_a - E_b)$ and $\sigma_{aa} + \sigma_{bb} = 1$. We can ignore the constant energy term $(E_a + E_b)/2$. We introduce the notation

$$\sigma_z = \sigma_{aa} - \sigma_{bb} = |a\rangle\langle a| - |b\rangle\langle b|, \tag{3.70}$$
$$\sigma_+ = \sigma_{ab} = |a\rangle\langle b|, \tag{3.71}$$
$$\sigma_- = \sigma_{ba} = |b\rangle\langle a|, \tag{3.72}$$

where σ_+ takes an atom in the ground state $|b\rangle$ to the excited state $|a\rangle$ and σ_- takes an atom from the excited state $|a\rangle$ to the ground state $|b\rangle$. Eq. (3.68) can be written in the form

$$H = \sum_k \hbar \nu_k a_k^\dagger a_k + \frac{1}{2}\hbar \omega \sigma_z + \hbar \sum_k g_k (\sigma_+ + \sigma_-)(a_k + a_k^\dagger). \tag{3.73}$$

This Hamiltonian contains four interacting terms. The term $a_k^\dagger \sigma_-$ describes a process where atom makes a transition from the upper to the lower energy state and a photon of mode k is created. The term $a_k \sigma_+$ is the opposite process. In both processes the energy is conserved. However, the term $a_k \sigma_-$ represents a process where atom makes transition from the upper to the lower level and a photon of mode k is absorbed. This leads to a loss of $2\hbar\omega$ energy. Similarly the term $a_k^\dagger \sigma_+$ represents a gain of $2\hbar\omega$ energy. We can ignore these two energy nonconserving terms. This approximation is called the *rotating-wave approximation*. The Hamiltonian after this approximation can be written as

$$H = \sum_k \hbar \nu_k a_k^\dagger a_k + \frac{1}{2}\hbar \omega \sigma_z + \hbar \sum_k g_k (\sigma_+ a_k + a_k^\dagger \sigma_-). \tag{3.74}$$

This Hamiltonian describes the interaction of a single two-level atom with a multi-mode radiation field. We can split this Hamiltonian in two parts as

$$H = H_0 + H_1, \tag{3.75}$$

where

$$H_0 = \sum_k \hbar \nu_k a_k^\dagger a_k + \frac{1}{2}\hbar\omega\sigma_z, \tag{3.76}$$

$$H_1 = \sum_k g_k(\sigma_+ a_k + a_k^\dagger \sigma_-). \tag{3.77}$$

The Hamiltonian (3.74) in the interaction picture is given as

$$H_I(t) = e^{iH_0 t/\hbar} H_1 e^{-iH_0 t/\hbar}. \tag{3.78}$$

We use the relation

$$e^{\alpha A} B e^{-\alpha A} = B + \alpha[A, B] + \frac{\alpha^2}{2!}[A, [A, B]] + \ldots, \tag{3.79}$$

to calculate the following relations

$$e^{i\nu_k a_k^\dagger a_k t} a_k e^{-i\nu_k a_k^\dagger a_k t} = a_k e^{-i\nu_k t}, \tag{3.80}$$

$$e^{i\omega\sigma_z t/2} \sigma_+ e^{-i\omega\sigma_z t/2} = \sigma_+ e^{i\omega t}. \tag{3.81}$$

Combining these relations, Eq. (3.78) can be given as

$$H_I(t) = \hbar \sum_k g_k \left[a_k^\dagger \sigma_- e^{-i(\omega-\nu_k)t} + a_k \sigma_+ e^{i(\omega-\nu_k)t} \right]. \tag{3.82}$$

Now our system corresponds to the two-level atom i.e., $\rho_S \equiv \rho_{atom}$. We insert the interaction energy $H_I(t)$ (Eq. (3.82)) into the equation of motion (3.59) and obtain

$$\begin{aligned}
\frac{d}{dt}\rho_{atom}(t) &= -i\sum_k g_k \langle a_k^\dagger \rangle [\sigma_-, \rho_{atom}(0)] e^{-i(\omega-\nu_k)t} - \int_0^t dx \sum_{k,k^x} g_k g_{k^x} \\
&\quad \Big\{ [\sigma_-\sigma_-\rho_{atom}(x) - 2\sigma_-\rho_{atom}(x)\sigma_- + \rho_{atom}(x)\sigma_-\sigma_-] \\
&\quad \times e^{-i(\omega-\nu_k)t - i(\omega-\nu_{k^x})x} \langle a_k^\dagger a_{k^x}^\dagger \rangle + [\sigma_-\sigma_+\rho_{atom}(x) - \sigma_+\rho_{atom}(x)\sigma_-] \\
&\quad \times e^{-i(\omega-\nu_k)t + i(\omega-\nu_{k^x})x} \langle a_k^\dagger a_{k^x} \rangle + [\sigma_+\sigma_-\rho_{atom}(x) - \sigma_-\rho_{atom}(x)\sigma_+] \\
&\quad \times e^{i(\omega-\nu_k)t - i(\omega-\nu_{k^x})x} \langle a_k a_{k^x}^\dagger \rangle \Big\} + \text{H.c.}, \tag{3.83}
\end{aligned}$$

where H.c. stands for the Hermitian conjugate and the expectation values refer to the initial state of the reservoir.

3.3.2 Atomic decay by thermal reservoirs

Let us take the reservoir variables in the uncorrelated thermal equilibrium mixture of states. The reduced density operator is the multi-mode extension of the thermal operator, given by

$$\rho_E = \prod_k \left[1 - \exp\left(-\frac{\hbar \nu_k}{k_\beta T}\right)\right] \exp\left(-\frac{\hbar \nu_k\, a_k^\dagger a_k}{k_\beta T}\right), \qquad (3.84)$$

where k_β is the Boltzmann constant and T is the temperature. It can be shown easily that

$$\langle a_k \rangle = \langle a_k^\dagger \rangle = 0, \qquad (3.85)$$
$$\langle a_k^\dagger a_{k^x} \rangle = \bar{n}_k\, \delta_{kk^x}, \qquad (3.86)$$
$$\langle a_k a_{k^x}^\dagger \rangle = (\bar{n}_k + 1)\, \delta_{kk^x}, \qquad (3.87)$$
$$\langle a_k a_{k^x} \rangle = \langle a_k^\dagger a_{k^x}^\dagger \rangle = 0, \qquad (3.88)$$

where $\bar{n}_k$ is the mean thermal photon number given by

$$\bar{n}_k = \frac{1}{\exp\left(\frac{\hbar \nu_k}{k_\beta T}\right) - 1}. \qquad (3.89)$$

We substitute these expectation values in Eq. (3.83) and get

$$\begin{aligned}\frac{d}{dt}\rho_{atom}(t) &= -\int_0^t dx \sum_k g_k^2 \Big\{ [\sigma_-\sigma_+\rho_{atom}(x) - \sigma_+\rho_{atom}(x)\sigma_-] \\ &\quad \times \bar{n}_k e^{-i(\omega-\nu_k)(t-x)} + [\sigma_+\sigma_-\rho_{atom}(x) - \sigma_-\rho_{atom}(x)\sigma_+] \\ &\quad \times (\bar{n}_k + 1)e^{i(\omega-\nu_k)(t-x)} \Big\} + \text{H.c.}\,, \end{aligned} \qquad (3.90)$$

We replace the sum over k by an integral

$$\sum_k \to 2\frac{V}{(2\pi)^3} \int_0^{2\pi} d\phi \int_0^\pi d\theta \sin\theta \int_0^\infty dk\, k^2, \qquad (3.91)$$

where V is the quantization volume. From Eq. (3.67) follows that

$$g_k^2 = \frac{\nu_k}{2\hbar\epsilon_0 V}\wp_{ab}^2 \cos^2\theta, \qquad (3.92)$$

where θ is the angle between the atomic dipole moment $\wp_{ab}$ and the electric field polarization vector $\hat{\epsilon}_k$. Substituting relations (3.91) and (3.92) in

Eq. (3.90), we get

$$\frac{d}{dt}\rho_{atom}(t) = -\frac{1}{4\pi\epsilon_0}\frac{4\wp_{ab}^2}{3\hbar c^3}\int_0^\infty d\nu_k \nu_k^3 \int_0^t dx\Big\{[\sigma_-\sigma_+\rho_{atom}(x) - \sigma_+\rho_{atom}(x)\sigma_-]$$
$$\times \bar{n}_k e^{-i(\omega-\nu_k)(t-x)} + [\sigma_+\sigma_-\rho_{atom}(x) - \sigma_-\rho_{atom}(x)\sigma_+]$$
$$\times (\bar{n}_k + 1) e^{i(\omega-\nu_k)(t-x)}\Big\} + \text{H.c.}, \tag{3.93}$$

where we have carried out integrations over θ and ϕ and used $k = \nu_k/c$. The intensity of light related with the emitted radiation is centred about the atomic transition frequency ω. The frequency ν_k^3 varies little around $\nu_k = \omega$ and the time integral for it in Eq. (3.93) can not be ignored. We can replace ν_k^3 by ω^3 and the lower limit in the ν_k integral by $-\infty$. With the integral

$$\int_{-\infty}^\infty d\nu_k \, e^{i(\omega-\nu_k)(t-x)} = 2\pi\,\delta(t-x), \tag{3.94}$$

Eq. (3.93) can be written by

$$\frac{d}{dt}\rho_{atom}(t) = -(\bar{n}_{th}+1)\frac{\Gamma}{2}[\sigma_+\sigma_-\rho_{atom}(t) - \sigma_-\rho_{atom}(t)\sigma_+]$$
$$-\bar{n}_{th}\frac{\Gamma}{2}[\sigma_-\sigma_+\rho_{atom}(t) - \sigma_+\rho_{atom}(t)\sigma_-] + \text{H.c.}, \tag{3.95}$$

where $\bar{n}_{th} \equiv \bar{n}_{k_0}(k_0 = \omega/c)$ and

$$\Gamma = \frac{1}{4\pi\epsilon_0}\frac{4\wp_{ab}^2}{3\hbar c^3} \tag{3.96}$$

is the atomic decay rate.

The equations of motion for the atomic density matrix are given by

$$\dot{\rho}_{aa} = \langle a|\dot{\rho}_{atom}|a\rangle$$
$$= -(\bar{n}_{th}+1)\Gamma\rho_{aa} + \bar{n}_{th}\Gamma\rho_{bb}, \tag{3.97}$$
$$\dot{\rho}_{ab} = \dot{\rho}_{ba}^* = -(\bar{n}_{th}+\frac{1}{2})\Gamma\rho_{ab} \tag{3.98}$$
$$\dot{\rho}_{bb} = (\bar{n}_{th}+1)\Gamma\rho_{aa} - \bar{n}_{th}\Gamma\rho_{bb}. \tag{3.99}$$

For the special case of vacuum reservoirs, i.e., the reservoir at zero-temperature ($\bar{n}_{th} = 0$), the equations of motion reduces to

$$\dot{\rho}_{aa} = -\Gamma\rho_{aa}, \tag{3.100}$$
$$\dot{\rho}_{ab} = -\frac{\Gamma}{2}\rho_{ab}, \tag{3.101}$$
$$\dot{\rho}_{bb} = \Gamma\rho_{aa}. \tag{3.102}$$

Chapter 4
Entanglement sudden death

We have argued in Chapter 1 that quantum entanglement is a resource for many applications of quantum information processing. For example, the fields of quantum computing [13, 14, 15, 16], quantum key distribution [17], and quantum teleportation [19, 21] all rely on having entangled states of at least two qubits. To accomplish the various quantum feats, the presence of entanglement among the parties sharing quantum states is both necessary and important. It was soon realized that entanglement is a dynamic resource and dynamics of entanglement depends on the choice of a physical system and an environment. The unavoidable interaction of entangled states with environments usually causes a decrement in the amount of entanglement. This unavoidable interaction is called decoherence. Decoherence is a serious limitation to quantum information processing. In addition, a deeper understanding of quantum decoherence is also necessary for bringing new insights into the foundations of quantum mechanics, in particular quantum measurement and quantum to classical transitions [58, 59, 60].

For entangled states with the subsystems coupled to their own individual environments, decoherence affects both the local and the global coherences. Since each qubit is inevitably subject to decoherence and decay processes, no matter how much they may be screened from the external environment, it is important to consider possible degradation of any initially established entanglement. It is no surprise that decoherence leads to a gradual decay (taking infinite time for complete disentanglement) of initially prepared entanglement. Yu and Eberly reported the surprising phenomenon that although local coherence is lost asymptotically there are some situations when global coherence (entanglement) is completely lost in a finite time [61]. Such finite-time disentanglement process is called *entanglement sudden death*. Yu and Eberly have studied a particular case where two initially entangled qubits are located inside two statistically independent vacuum reservoirs.

They showed that the simple phenomenon of spontaneous emission caused by vacuum fluctuations have different effect on local and global coherences. Some quantum states undergo finite-time disentanglement while others do not. Soon after this study, Jakóbczyk and Jamróz showed that certain entangled states of two qubits interacting with two independent thermal baths at very high temperatures exhibit finite-time disentanglement [62]. Dodd and Halliwell showed the existence of sudden death for continuous-variable systems [58, 59]. Later, Yu and Eberly demonstrated this effect under classical noise [63]. Below we demonstrate this phenomenon both in amplitude damping and in phase damping environments. Experimental evidences for this effect have been reported recently [64, 65, 66]. Clearly, such finite-time disappearance of entanglement can seriously affect its applications in quantum information processing.

This chapter is organized as follows. In Section 4.1, we describe sudden death caused by amplitude damping. In Section 4.2, we discuss sudden death via phase damping. Other investigations regarding finite-time disentanglement are discussed in Section 4.3.

4.1 Sudden death via amplitude damping

The contents of this section are taken from the original work of Yu and Eberly [61]. We briefly reproduce the key results here. We consider two two-level atoms A and B coupled individually to two environments which are initially in their vacuum states (compare with Figure 4.1. The two

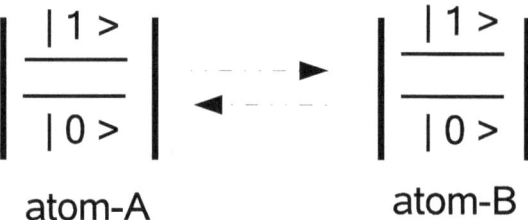

Figure 4.1: Two atoms located inside two statistically independent vacuum reservoirs: The atoms are initially entangled but are not interacting directly.

atoms are initially prepared in an entangled state and they both interact with their local environments. There is no direct interaction between the atoms anymore. The interaction between each atom and its environment leads to a loss of both local coherence and quantum entanglement of the two atoms. The total Hamiltonian of the combined system is given by

$$H_{tot} = H_{at} + H_{env} + H_{int}, \qquad (4.1)$$

where H_{at} is the Hamiltonian of the two atoms, H_{env} is the Hamiltonian of the environments, and H_{int} is the Hamiltonian describing the interaction between the atoms and the reservoirs. The standard two-qubit product basis is denoted $|1\rangle_{AB} = |1,1\rangle_{AB}$, $|2\rangle_{AB} = |1,0\rangle_{AB}$, $|3\rangle_{AB} = |0,1\rangle_{AB}$, and $|4\rangle_{AB} = |0,0\rangle_{AB}$, where $|1,1\rangle_{AB}$ describes the state in which both atoms are in their excited states, etc.

Following the standard methods of averaging over reservoirs degrees of freedom, using the dipole- and rotating-wave approximations, and solving the master equation, one obtains the complete dynamics of system of the two atoms (see Chapter 3). We demonstrate finite-time disentanglement in this particular setup by choosing an initially entangled state

$$\rho = \frac{1}{3}\begin{pmatrix} a & 0 & 0 & 0 \\ 0 & 1 & 1 & 0 \\ 0 & 1 & 1 & 0 \\ 0 & 0 & 0 & 1-a \end{pmatrix}, \qquad (4.2)$$

with $0 \leq a \leq 1$. The time evolution of this quantum state can be determined completely from the most general solution provided in Chapter 5 by Eqs. (5.7-5.16). After obtaining the time dependent density matrix $\rho(t)$, it is simple to calculate *concurrence*. For the density matrix $\rho(t)$ it is given by [61]

$$C(\rho(t)) = \frac{2}{3}\max\{0, \gamma^2 f(t)\}, \qquad (4.3)$$

where $\gamma = \exp[-\Gamma t/2]$, and $f(t) = 1 - \sqrt{a(1-a+2\omega^2+\omega^4 a)}$ with $\omega = \sqrt{1-\exp[-\Gamma t]}$. This concurrence is plotted in Figure 4.2, which shows that for all values of a between $1/3$ and 1, concurrence decay is completed in a finite time, but for smaller values of a, the decay time is infinite. For the special case of $a = 1$, $C(\rho(t)) = 0$ for all $t \geq t_d$, where t_d is finite and given by

$$t_d \equiv \frac{1}{\Gamma}\ln\left[\frac{2+\sqrt{2}}{2}\right]. \qquad (4.4)$$

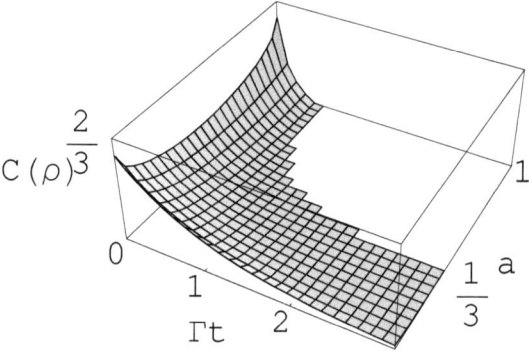

Figure 4.2: *Concurrence* is plotted against the decay parameter Γt and the single parameter a: Finite-time disentanglement takes place for $a > 1/3$, whereas for $a \leq 1/3$, entanglement decays only asymptotically.

4.2 Sudden death via phase damping

Phase damping or classical noise is another type of decoherence responsible for decay of both local and global coherences. In this section, we describe the possibility of entanglement sudden death arising from the influence of classical noise on two qubits which are initially prepared in an entangled state but have no direct interaction. The discussion of decoherence due to classical noise can be further divided into two classes. In Section 4.2.1, we describe the effects of global collective noise on both qubits. The effects of exposing each qubit separately to local noise are discussed in Section 4.2.2.

4.2.1 Disentanglement due to global collective noise

We consider two qubits initially prepared in an entangled state which are affected collectively by a single stochastic field. The Hamiltonian of the qubits plus the classical noisy field is given by

$$H(t) = -\frac{1}{2} \mu\, B(t)\, (\sigma_z^A + \sigma_z^B), \tag{4.5}$$

where μ is the gyromagnetic ratio, and $\sigma_z^{A,B}$ are the Pauli matrices in the standard basis defined in Section 4.1. We assume that $B(t)$ is a Gaussian field and satisfies the Markov condition

$$\begin{aligned} \langle B(t) \rangle &= 0, \\ \langle B(t)\, B(t') \rangle &= \frac{\Gamma}{\mu^2}\, \delta(t-t'), \end{aligned} \qquad (4.6)$$

where $\langle \ldots \rangle$ stands for an ensemble average and Γ is the dephasing damping rate due to the collective interaction with $B(t)$.

The solution for the reduced system under the Hamiltonian (4.5) can be obtained by various methods, e. g. master equation, stochastic Schrödinger equation, and the operator sum representation. The reduced density matrix for the two qubits can be obtained from the statistical density operator $\rho_{st}(t)$ for both qubits and a classical Gaussian field by taking the ensemble average over the noisy field $B(t)$ given by

$$\rho(t) = \langle \rho_{st}(t) \rangle, \qquad (4.7)$$

where the statistical density operator $\rho_{st}(t)$ is given by

$$\rho_{st}(t) = U(t)\, \rho(0)\, U^\dagger(t), \qquad (4.8)$$

with the unitary operator $U(t) = \exp[-\mathrm{i} \int_0^t dt' H(t')]$. The explicit form of the unitary operator is given by

$$U(t) = \exp[\,\mathrm{i}\, \frac{\mu}{2} \int_0^t dt'\, B(t')\, (\sigma_z^A + \sigma_z^B)\,]. \qquad (4.9)$$

We can average over noise degrees of freedom in Eq. (4.8) and can write the most general solution in terms of the Kraus operators [67]

$$\rho(t) = \sum_{j=1}^{3} K_j^\dagger(t)\, \rho(0)\, K_j(t), \qquad (4.10)$$

where the Kraus operators describing the collective interaction are given by

$$K_1 = \begin{pmatrix} \gamma & 0 & 0 & 0 \\ 0 & 1 & 0 & 0 \\ 0 & 0 & 1 & 0 \\ 0 & 0 & 0 & \gamma \end{pmatrix},$$

$$K_2 = \begin{pmatrix} \omega_1 & 0 & 0 & 0 \\ 0 & 0 & 0 & 0 \\ 0 & 0 & 0 & 0 \\ 0 & 0 & 0 & \omega_2 \end{pmatrix},$$

$$K_3 = \begin{pmatrix} 0 & 0 & 0 & 0 \\ 0 & 0 & 0 & 0 \\ 0 & 0 & 0 & 0 \\ 0 & 0 & 0 & \omega_3 \end{pmatrix}, \qquad (4.11)$$

where $\gamma = e^{-\Gamma t/2}$, $\omega_1 = \sqrt{1-\gamma^2}$, $\omega_2 = -\gamma^2\sqrt{1-\gamma^2}$, $\omega_3 = (1-\gamma^2)\sqrt{1+\gamma^2}$.

Let us consider a special class of mixed states namely X-states, where the only non-zero matrix elements are on diagonal and anti-diagonal positions. The density matrix for X-states is given by

$$\rho_X = \begin{pmatrix} \rho_{11} & 0 & 0 & \rho_{14} \\ 0 & \rho_{22} & \rho_{23} & 0 \\ 0 & \rho_{32} & \rho_{33} & 0 \\ \rho_{41} & 0 & 0 & \rho_{44} \end{pmatrix}. \qquad (4.12)$$

Eq. (4.10) leads to

$$\rho_X(t) = \begin{pmatrix} \rho_{11} & 0 & 0 & \gamma^4\rho_{14} \\ 0 & \rho_{22} & \rho_{23} & 0 \\ 0 & \rho_{32} & \rho_{33} & 0 \\ \gamma^4\rho_{41} & 0 & 0 & \rho_{44} \end{pmatrix}. \qquad (4.13)$$

From Eq. (4.13), it is clear that the collective noise only affects the off-diagonal elements ρ_{14} and ρ_{41} and leaves all other elements intact. In particular the diagonal elements remain constant in time. This is in contrast to amplitude damping, where all matrix elements are affected. For pure phase damping, the collective global field allows certain phase combinations to cancel out and generates a decoherence-free subspace [68, 69] spanned by $|1,0\rangle$ and $|0,1\rangle$. However, we avoid such protection by assuming $\rho_{23} = \rho_{32} = 0$. Concurrence of $\rho_X(t)$ is given by

$$C(\rho_X(t)) = 2\max\left\{0, |\rho_{14}(t)| - \sqrt{\rho_{22}\rho_{33}}, |\rho_{23}(t)| - \sqrt{\rho_{11}\rho_{44}}\right\}. \qquad (4.14)$$

Therefore $\rho_X(t)$ is separable if and only if $|\rho_{14}(t)| - \sqrt{\rho_{22}\rho_{33}} \leq 0$, and $|\rho_{23}(t)| - \sqrt{\rho_{11}\rho_{44}} \leq 0$. Concurrence of the density matrix (4.13) with $\rho_{23} = \rho_{32} = 0$ is given as

$$C(\rho_X(t)) = 2\max\left\{0, |\rho_{14}|e^{-2\Gamma t} - \sqrt{\rho_{22}\rho_{33}}\right\}. \qquad (4.15)$$

The critical time for finite-time disentanglement is given as

$$t_c = \frac{1}{2\Gamma}\ln\frac{|\rho_{14}|}{\sqrt{\rho_{22}\rho_{33}}} \qquad (4.16)$$

with $C(\rho_X(t)) = 0$ for $t \geq t_c$. It is clear from Eq. (4.16) that for $\rho_{22} \neq 0$ and $\rho_{33} \neq 0$, sudden death will occur at time t_c. If any of these matrix elements is zero, then entanglement decays asymptotically. These features are shown in Figure 4.3.

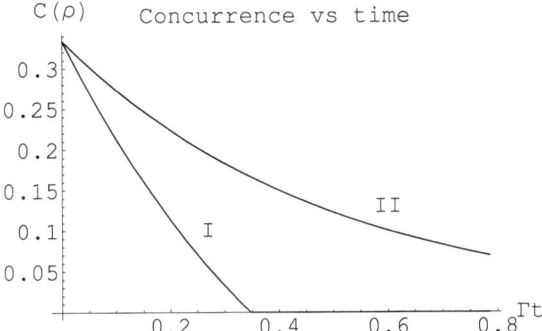

Figure 4.3: *Concurrence* is plotted against the decay parameter Γt: The graph shows sudden death process (I) with initial values $\rho_{23} = 0$, $\rho_{14} = 1/3$, $\rho_{11} = \rho_{44} = 1/3$, $\rho_{22} = \rho_{33} = 1/6$ and the exponential decay (II) with $\rho_{23} = 0$, $\rho_{14} = 1/6$, $\rho_{11} = \rho_{44} = \rho_{33} = 1/3$, $\rho_{22} = 0$. There is no sudden death for mixed states with $\rho_{22} = 0$ or $\rho_{33} = 0$.

4.2.2 Disentanglement due to local noise

In this situation, two qubits which may be initially prepared in an entangled state, interact with their own local environment represented by two independent classical noise sources. The Hamiltonian of two qubits plus classical local noises is given as

$$H(t) = -\frac{1}{2}\mu \left(b_A(t)\,\sigma^A + b_B(t)\,\sigma^B\right), \tag{4.17}$$

where the noise sources $b_A(t)$ and $b_B(t)$ are statistically independent classical Gaussian field and satisfy the Markov condition

$$\langle b_i(t) \rangle = 0,$$
$$\langle b_i(t)\,b_i(t') \rangle = \frac{\Gamma_i}{\mu^2}\delta(t-t'), \quad i = A, B, \tag{4.18}$$

where Γ_i are the phase damping rates of qubits due to coupling to the stochastic magnetic fields $b_i(t)$.

Similar to the case of collective noise, the general solution of the reduced density operator of two qubits can be written as [63]

$$\rho(t) = \sum_{u,v=1}^{2} E_u^\dagger(t)\,F_v^\dagger(t)\,\rho(0)\,F_v(t)\,E_u(t), \tag{4.19}$$

where the Kraus operators describing the interaction with local environments are given by

$$E_1 = \begin{pmatrix} 1 & 0 \\ 0 & \gamma_A \end{pmatrix} \otimes I, \quad E_2 = \begin{pmatrix} 0 & 0 \\ 0 & \omega_A \end{pmatrix} \otimes I, \quad (4.20)$$

$$F_1 = I \otimes \begin{pmatrix} 1 & 0 \\ 0 & \gamma_B \end{pmatrix}, \quad F_2 = I \otimes \begin{pmatrix} 0 & 0 \\ 0 & \omega_B \end{pmatrix}, \quad (4.21)$$

with $\gamma_A = e^{-\Gamma_A t/2}$, $\gamma_B = e^{-\Gamma_B t/2}$, $\omega_A = \sqrt{1-\gamma_A^2}$, $\omega_B = \sqrt{1-\gamma_B^2}$.

For X-states (4.12), the general solution is given by

$$\rho(t) = \begin{pmatrix} \rho_{11} & 0 & 0 & \gamma_A\gamma_B\rho_{14} \\ 0 & \rho_{22} & \gamma_A\gamma_B\rho_{23} & 0 \\ 0 & \gamma_A\gamma_B\rho_{32} & \rho_{33} & 0 \\ \gamma_A\gamma_B\rho_{41} & 0 & 0 & \rho_{44} \end{pmatrix}. \quad (4.22)$$

This means that when both qubits are exposed to local dephasing separately, there is no decoherence-free subspace. Comparing Eq. (4.22) with Eq. (4.13), we observe that sudden death may appear in this case as well. However, the time of disentanglement is different due to the change of noise but sudden death appears for all non-zero diagonal elements.

4.3 Further recent investigations

In this section, we shortly review recent results of investigations on sudden death of entanglement in various situations. The work of Yu-Eberly [61, 63] and Jakóbczyk-Jamróz [62] attracted a lot of people to this problem. Dodd-Halliwell [58, 59] investigated the process of disentanglement for the continuous variables systems. The dynamics of two-qubits entanglement was explored in symmetry-broken environment [70]. It was shown [71] that for pure decoherence the decay of two-qubit entanglement is approximately governed by the product of the suppression factors describing docoherence of subsystems, if they are subjected to uncorrelated noise. Liang showed that if the initial state is not a maximally entangled state then entanglement decays faster than the product of the suppression factors describing decoherence of qubits [72]. Entanglement sudden death of two-qubits in Jaynes-Cummings model was investigated as well [73]. The time evolution of entanglement for bipartite systems of arbitrary dimensions was also investigated [74]. The analysis of sudden death of two-qubit X-states under amplitude damping, phase damping and state-equalizing noise was done in [75]. It was also shown that the partial dephasing induced by a super-Ohmic reservoir, may also lead to sudden death [76]. The phase-induced collapse and revival of entanglement of two-qubit entangled states interacting in a trap was predicted [77]. Ban investigated decoherence of the Gaussian states under the influence of non-Markovian quantum channels [78] and the correlated and collective stochastic dephasing of two-qubit entanglement [79]. The direct measurement of ESD was proposed [80] through the measurement of a single observable invariant with respect to decay process. This was an additional effort to give physical meaning to measures of entanglement. Decoherence of entanglement of two-qubits interacting via a Heisenberg XY chain interaction was studied in [81]. Lamata et al. showed that entanglement between two-qubits decreases when the correlations are transferred locally to the momentum degree of freedom of one of the qubit [82]. Jamróz studied local aspects of ESD induced by spontaneous emission and showed that locally equivalent entangled states exhibit different behavior in their disentanglement process [83]. Yu-Eberly demonstrated another surprising result that if a single qubit is exposed to both amplitude damping and phase damping, the decay rate is additive, however for the simplest case of entangled two-qubits this additivity of decay rate breaks down [84]. Ficek and Tanas showed that when two qubits are coupled collectively to a multimode field, the irreversible spontaneous decay can lead to a revival of entanglement that has already been destroyed [85]. The effect of quantum interference [86] on entanglement of two three level atoms has been studied [87]. It was shown that quantum interference can slow down the process of disentanglement and for maximum interference this system (qutrit-qutrit) has non-trivial asymptotic entangled states.

Cui et al. studied ESD for bipartite systems subjected to different scenarios [88]. Sun et al. investigated the dynamics of entanglement for two-qubits and two-qutrits coupled to an Ising spin chain in a transverse field [89]. Huang and Zhu worked out the necessary and sufficient conditions for sudden death of entanglement of two-qubits via phase damping and amplitude damping [90]. Ann and Jaeger demonstrated finite-time disentanglement due to multi-local dephasing noise for a class of bipartite states in finite-dimensional systems [91]. We studied disentanglement due to amplitude damping in qubit-qutrit systems [92] (see Section 6.1). Ann and Jaeger studied finite-time disentanglement for qubit-qutrit systems due to phase damping [93]. Ikram et al. studied the time evolution of various entangled states of two-qubits exposed to thermal and squeezed reservoirs. They numerically showed that maximally entangled state (Bell state) exhibits sudden death interacting with independent thermal reservoirs and conjectured that all quantum states exhibit sudden death in thermal reservoirs [94]. Cunha discussed the geometrical point of view for sudden death [95]. Eberly-Yu mentioned the implications of sudden death for quantum information processing [96]. Lastra et al. investigated the abrupt changes in disentanglement in qutrit-qutrit systems [97]. The story took an important turn by the experimental verification of sudden death. Sudden death has been reported in laboratory for optical setups [64, 65] and for an atomic ensemble [66]. Fanchini and Napolitano studied protection of sudden death of two-qubits using continuous dynamical decoupling [98].

Another parallel investigation focused on non-locality of quantum states via violation of Bell inequalities. The quantum states which initially violates a Bell inequality may not violate it after some finite time. Such studies have been extensively done for different scenarios, see for example Refs. [99, 100, 101] and references therein.

We have extended the discussion of sudden death towards its hastening, delaying, and avoiding. We have shown that simple local unitary actions can delay or even completely avoid sudden death of two-qubits interacting with independent reservoirs at zero-temperature [102]. Al-Qasimi and James studied disentanglement of two-qubits interacting with independent reservoirs at finite-temperature. Similar to our analytical conclusions [129], they showed that large number of quantum states exhibit sudden death at finite-temperature [103]. Bellomo et al. studied entanglement dynamics of two-qubits interacting with independent reservoirs at zero-temperature with non-Markovian assumption. They also considered the effect of two-qubits entanglement dynamics interacting with statistically independent finite-temperature reservoirs without memory [104]. Cao and Zheng investigated disentanglement process for certain classes of two-qubits entangled states in non-Markovian approach. They have shown that entanglement decays asymptotically only in the case of weak coupling between a system and an environment. For strong coupling between a qubit and an environment, entanglement sudden death always appear even if the reservoirs are at zero-temperature [105]. Lastra et al. studied the time evolution of a class of entangled coherent states of two electromagnetic fields under dissipation. They concluded that asymptotic decay and sudden death can be found depending on the initial condition and phase space distances among the component of each mode [106]. Hernandes and Orszag studied disentanglement for two qubits in a common squeezed reservoir and worked out the relation between sudden death and revival of entanglement [107]. Sainz and Björk suggested a scheme to delay, cause and combat sudden death with non-local quantum error correction [108, 109]. Tahira et al. investigated entanglement dynamics of a pure bipartite system subjected to dissipative environments. They found that doubly excited component (means both atoms are in excited states) is a necessary condition for sudden death in vacuum reservoirs, while all pure states exhibit sudden death in finite-temperature reservoirs [110]. Derkacz and Jakóbczyk studied the dynamics of entanglement between two three-level atoms coupled to the common vacuum reservoir [111]. Jian Li *et al.* analyzed the enhancement of sudden death for driven qubits and shown that the time of sudden death decreases due to driving [112]. Wang *et al.* explored decoherence of two-qubits in a non-Markovian reservoir without rotating wave approximation [113]. Yamamoto *et al.* have shown that sudden death appearing in a linear quantum network can be avoided via direct measurement feedback method [114]. Scala *et al.* analyzed entanglement dynamics of two coupled qubits while interacting with two independent bosonic baths [115]. Shan *et al.* investigated entanglement properties for two-qubits in the Heisenberg XY interaction and subjected to a magnetic field. They found that sudden death and sudden birth of entanglement appear during the evolution process for particular initial states [116]. Peng Li *et al.* shown that for composite noise environment, two-qubits Hilbert space can be divided into two parts: a 3-dimensional subspace in which all states disentangle asymptotically and that in which all states exhibit finite-time disentanglement [?]. Cole observed that sudden death of entanglement appears when bipartite entanglement is shared among multiparties. Hence for a particular partition of this multipartite entangled system, ESD may appear [118]. Chan *et al.* investigated entanglement dynamics of two distant but non-identical atoms interacting independently a cavity field as in the Jaynes-Cummings model [119]. Dajka and Luczka studied the origination and survival of qudit-qudit (two d-dimensional systems) entanglement for various environments [120]. Marek *et al.* derived a sufficient condition for infinite dimensional systems which do not exhibit sudden death while interacting with vacuum environments. They clarified a class of bipartite entangled states which are tolerant against decoherence in a vacuum [121]. Drumond and Cunha analyzed entanglement dynamics of two-qubits interacting with several relaxing environments in the light of geometry [122]. Dubi and Ventra studied two interacting qubits in a magnetic field and a thermal Markovian environment, presenting nonmonotonic relaxation rates as functions of the magnetic field and temperature [123]. Al-Qasimi and James showed that for the continuous variable quantum systems, two-mode-N-photon states undergoing pure dephasing never exhibit sudden death [124]. Lastra *et al.* studied disentanglement of two distinguishable atoms and quantum recoil effects [125]. Paz and Roncaglia studied two harmonic oscillators coupled to a common environment also modelled as oscillators [126]. Sudden death of entanglement and entanglement revival are examined as well as non-Markovian environments. López *et al.* found that for two-qubits interacting with independent reservoirs, sudden death of entanglement necessarily means sudden birth of entanglement (ESB) in reservoirs [127]. Mazzola *et al.* studied entanglement dynamics of two-qubits in a common structured reservoir. They have shown that both sudden death and sudden birth of entanglement appear for certain states [128].

Quite recently, we have analyzed entanglement dynamics of a complete set of two-qubit X-states interacting with vacuum and thermal reservoirs (Section 5.2). We have also demonstrated hastening, delaying, and avoiding sudden death of entanglement in zero- and finite-temperature reservoirs [129]. Yu-Eberly have reviewed the recent progress on sudden death [130]. An and Kim studied entanglement dynamics of multipartite Greenberger-Horne-Zeilinger-type (GHZ) states exposed to phase damping and amplitude damping simultaneously. They have reported the existence of sudden death for such states, however local operations can circumvent such finite-time disentanglement [131]. More recently, Ann and Jaeger have reviewed theoretical and experimental work for sudden death of entanglement [132].

Chapter 5

Hastening, delaying or avoiding entanglement sudden death of qubit-qubit systems

We have mentioned in Chapter 1 that several applications of quantum information rely on entangled states of two-qubits. In this Chapter we address the problem of sudden death of qubit-qubit systems. In particular, we want to prevent this unwanted phenomenon. Sudden death of entanglement has been investigated initially for two-qubits theoretically [61, 63]. Experimental evidences of this phenomenon have been provided only recently [64, 65, 66].

It is well known in the context of spontaneous emission processes [133, 134] or delayed choice experiments [135], for example, that characteristic quantum phenomena and effects of decoherence can be influenced significantly by measurements. Therefore, it would be of interest if Alice and Bob, the two holders of an entangled pair, can take suitable individual actions to postpone sudden death of entanglement. Some preliminary studies on changing the initial state into an equivalent more robust entangled state have been carried out recently [73, 75, 83].

In this Chapter, we deal with the question that, even given an initial state and a setup which exhibits finite-time disentanglement, can we take suitable actions later to change the dynamics of entanglement? We answer in the affirmative. In particular, we show that simple local unitary operations can change the time of sudden death. This is even possible if these local operations are separated in space so that they are space-like and cannot be connected by any causal relation. The operations we consider can either hasten or delay that time depending on their time of application. A suitable window for this application can even avert completely finite-time disentanglement. In that case, entanglement will persist and decay only asymptotically. Similar results will apply also to other systems such as qubit-qutrit [92, 100] and qutrit-qutrit systems [87, 97] where also questions of the finite end of entanglement have been considered. We will discuss these results in detail in Chapter 6.

In this Chapter we consider entangled states of two-qubits interacting with statistically independent vacuum and thermal reservoirs [102, 129]. In Section 5.1, we study hastening, delaying or avoiding sudden death of two entangled qubits interacting with statistically independent vacuum reservoirs. In Section 5.2, we discuss our physical model and the master equation along with its solution. In Section 5.3, we present analytical results for hastening, delaying, or avoiding sudden death of two qubits entangled states interacting with statistically uncorrelated zero-temperature reservoirs. Analytical results for the appearance of sudden death valid for so called X-states are provided in Section 5.4. We also provide numerical evidence for the possibility of hastening, and delaying sudden death in finite-temperature reservoirs.

5.1 Numerical evidence for hastening, delaying or avoiding sudden death

We consider a model consisting of two two-level atoms each interacting with its own independent reservoir. We take both reservoirs to be in their ground states (compare with Figure 5.4). "Amplitude damping" in the form of spontaneous (pure exponential) decay into statistically independent reservoirs from the excited to the ground state is the particular dynamics we want to investigate [61, 68, 69, 96]. In this special case, the final state is a product state of the two ground-state qubits with no entanglement. It is a pure state with zero entropy. Let us restrict ourselves in the subsequent discussion to mixed initial states with a density matrix of the form

$$\rho(t) = \frac{1}{3} \begin{pmatrix} a(t) & 0 & 0 & 0 \\ 0 & b(t) & z(t) & 0 \\ 0 & z(t) & c(t) & 0 \\ 0 & 0 & 0 & d(t) \end{pmatrix}. \tag{5.1}$$

The coefficients (a, b, c, d, z) are real-valued and non-negative except z, which may possibly be negative and $\text{tr}(\rho) = (a + b + c + d)/3 = 1$. The four orthonormal basis states of two-qubits are denoted by $(|1,1\rangle, |1,0\rangle, |0,1\rangle, |0,0\rangle)$ and $|1\rangle/|0\rangle$ denotes the excited/ground state.

The states in Eq. (5.1) are of interest because they maintain their basic structure during interaction with independent reservoirs as shown in Section 5.2. This implies that all off-diagonal density matrix elements which are absent at $t = 0$, remain zero throughout the dynamical evolution. The dynamics of this particular model is governed by the master equation (5.6) with $m = n = 0$. Let Γ be the spontaneous decay rate of both atoms, i.e., $\gamma_1 = \gamma_2 = \Gamma$ in Eq. (5.6). The initial condition chosen, of $b(0) = c(0) = z(0) = 1$, along with the only evolution, that $|1\rangle$ decays to $|0\rangle$ at a steady rate $\exp(-\Gamma t/2)$ in amplitude, keeps $b(t) = c(t)$ throughout.

The off-diagonal density matrix element $z(t)$ is a solution of the differential equation $\dot{z}(t) = -\Gamma z(t)$ (an overhead dot indicates differentiation with respect to time). The diagonal elements are obtained as

$$\frac{d}{dt}\begin{pmatrix} \rho_{11} \\ \rho_{22} \\ \rho_{33} \\ \rho_{44} \end{pmatrix} = \begin{pmatrix} -2\Gamma & 0 & 0 & 0 \\ \Gamma & -\Gamma & 0 & 0 \\ \Gamma & 0 & -\Gamma & 0 \\ 0 & \Gamma & \Gamma & 0 \end{pmatrix} \begin{pmatrix} \rho_{11} \\ \rho_{22} \\ \rho_{33} \\ \rho_{44} \end{pmatrix}. \quad (5.2)$$

These equations have an obvious structure dictated by the "decay" from the excited state $|1\rangle$ to "feed" into the ground state $|0\rangle$. Their solutions in terms of a logarithmic, dimensionless time parameter, $\gamma = \exp(-\Gamma t/2)$, are given as

$$\begin{aligned} \rho_{11}(t) = a(t) &= a(0)\gamma^4 \\ \rho_{22}(t) = \rho_{33}(t) = b(t) &= [b(0) + a(0)]\gamma^2 - a(0)\gamma^4 \\ \rho_{44}(t) = d(t) &= 3 + a(0)(\gamma^4 - \gamma^2) - [3 - d(0)]\gamma^2 \\ z(t) &= z(0)\gamma^2. \end{aligned} \quad (5.3)$$

Were (b, c, z) the only non-zero elements in Eq. (5.1), we would have an entangled pure state of the form $1/\sqrt{2}(|1,0\rangle + |0,1\rangle)$. The coefficients would all decay with a factor γ^2, and entanglement would decay only asymptotically. A choice of either $(a(0) = 1, d(0) = 0)$ or $(a(0) = 0, d(0) = 1)$ yields a mixed state which is non-separable. Both choices give the same Von Neumann entropy of magnitude $\ln(3/4^{1/3})$ at $t = 0$. It decreases to zero asymptotically when the system is in the pure state $|0,0\rangle$. Nevertheless, their evolution of entanglement is very different [61, 62, 68, 69, 80, 94, 96, 103].

The second choice of $d(0) = 1$ leads to non-separability only asymptotically at infinite times t, whereas the first choice with $a(0) = 1$ leads to a finite-time end of entanglement, i.e., "sudden death" [61, 68, 69, 96]. We choose *negativity* [37, 38] as an indicator of non-separability. For $2 \otimes 2$ systems, there can be at most one possible negative eigenvalue [136]. After taking the partial transpose of Eq. (5.1), the only possible negative eigenvalue is given by $\left[a(t) + d(t) - \sqrt{[a(t) - d(t)]^2 + 4z^2(t)}\right]/6$. This value can take negative values as long as $a(t)\,d(t) < z^2(t)$. When $a(0) = 0$, $a(t)$ from Eq. (5.3) remains zero at all times. Thus the system retains non-separability for all finite times t. However, for the choice $(d(0) = 0, a(0) = 1)$, the system starts as non-separable or entangled but when $a(t)d(t)$ becomes larger than $z^2(t)$ during the subsequent evolution, entanglement is lost. It can be seen that the time t_0 at which this happens is given as

$$t_0 = \frac{1}{\Gamma}\ln(1/\gamma_0^2) = \frac{1}{\Gamma}\ln(1 + 1/\sqrt{2}). \tag{5.4}$$

This is the time of "sudden death" [61, 68, 69, 96]. At this point, we have $a = 6 - 4\sqrt{2}$, $b = 2\sqrt{2} - 2$, $d = 1$, $z = 2 - \sqrt{2}$.

Previous research has examined the evolution of entanglement for different initial choices of the above parameters. The evolution of Werner states [36] has also been studied [61, 62, 68, 69, 80, 94, 96, 103]. It has been noted that different "initializations", wherein an initial given state such as in Eq. (5.1) is switched to another state with equivalent entanglement, can lead to a change in the time of non-separability [73, 75, 83]. Recent work has established necessary and sufficient conditions for this phenomenon under both amplitude and phase damping [90].

Another observation is that spontaneous emission can also lead to a revival of entanglement from a separable configuration [85]. In three-level atoms or qutrits, finite-end of entanglement for pairs and abrupt changes in lower bounds on entanglement have been noted [97]. For qutrit-qutrit systems, it was noted that quantum interference between different excited levels may create an asymptotic entanglement [87]. Sudden death of entanglement has been noted for mixed qubit-qutrit states [92, 100] as well and it seems to be a generic phenomenon for all entangled-pair systems.

We turn, however, to a different question. Given the qubit-qubit system above and the initial conditions that lead to separability in finite time, can a suitable intervention alter this time? This question is of practical interest because, as noted in [76], "finite-end of entanglement may affect the feasibility of solid-state based quantum computing". Therefore, a simple intervention that prolongs entanglement in quantum states can be of broad interest. Indeed, the above discussion and, especially, the two choices of whether it is a or d that is initially zero in Eq. (5.1), suggests a way for such an intervention. If $a(t) = 0$, non-separability of the mixed state because of the presence of $d(t)$ simply continues as the states that are entangled "decay down" to enhance $d(t)$. The other situation of $a(0) = 1$ and, therefore, a non-zero $a(t)$ is quite different, because it feeds into entangled sector "from above". At a crucial time when $a(t)\,d(t) > z^2(t)$, the mixed state becomes separable.

The above diagnosis of which aspect of the asymmetry between the parameters a and d of Eq. (5.1) is responsible for separability suggests a "switch" between them. Such a switch, which leaves the other coefficients (b, c, z) unchanged, amounts to interchanging states $|1\rangle$ and $|0\rangle$ for both qubits. This is a local unitary transformation that both Alice and Bob can easily implement, by individual σ_x operations for spins or by laser coupling of the excited and the ground states for two-level atoms. Consider the same initial condition as before, with $(a(0) = 1, d(0) = 0)$, which leads to sudden death. Before time t_0 corresponding to the end of entanglement, consider such local unitary operations that merely interchange the parameters a and d of Eq. (5.1). If this is done at the time t_A when $a = d$, which happens when $\exp(-\Gamma t_A) \equiv \gamma_A^2 = 3/4$, clearly there will be no effect upon the subsequent evolution, the end still coming at time t_0 (See Figure 5.1 and Figure 5.2). If the switch is made at any time intermediate between t_A and t_0 (see Figure 5.2), separability occurs earlier, a minimum being at the switch time t_1 with $t_1 \approx 0.357/\Gamma$.

More interestingly, a switch earlier than t_A prolongs entanglement as shown in the Figures. Moreover, switch times before t_B with $t_B \approx 0.1293/\Gamma$ avoid the finite time end all together, leading to separability only asymptotically. As a practical matter, therefore, Alice and Bob can make local unitary switch between states $|1\rangle$ and $|0\rangle$ at a certain time as desired to alter that end.

In Figure 5.1, *negativity* is plotted against the parameter Γt. The solid line corresponds to a situation when no switch is made and sudden death appears at time t_0 with $t_0 \approx 0.5348/\Gamma$. If the switch is made at time t_1 with $t_1 \approx 0.357/\Gamma$ (dotted-dashed line), sudden death reaches a minimum value of time $t_< \approx 0.48/\Gamma$. Any switch made earlier than t_A leads to a delay in sudden death in comparison with t_0. Two instances are shown in the upper curves. If the switch is made at time t with $t \approx 0.223/\Gamma$ (dashed line), *negativity* comes to an end at $\Gamma t \approx 0.716$. Any switch made earlier than time t_B with $t_B \approx 0.1293/\Gamma$ avoids a finite end, leading only to asymptotic decay of entanglement.

Figure 5.2 displays the time of sudden death t_{end} against the time of switching t_{sw}. The earlier the switch is made than time t_A with $t_A \approx 0.2877/\Gamma$, the more the end of entanglement is delayed. Sudden death is avoided completely when the switch takes place earlier than time t_B with $t_B \approx 0.1293/\Gamma$.

Interestingly, switching states $|1\rangle$ and $|0\rangle$ at only one end, that is, either

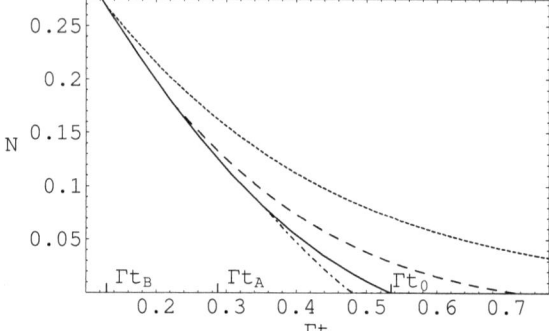

Figure 5.1: Evolution of *negativity* N for an initial mixed state in Eq. (5.1) with $a(0) = b(0) = c(0) = z(0) = 1, d(0) = 0$. The solid line shows the undisturbed evolution, with sudden death at $t_0 \approx 0.5348/\Gamma$. Other dotted and dashed lines show the effect of switching the values of a and d at different times: Those after time $t_A \approx 0.2877/\Gamma$ hasten, and those before delay, sudden death; switches earlier than $t_B \approx 0.1293/\Gamma$ avoid sudden death altogether with *negativity* vanishing only asymptotically in time.

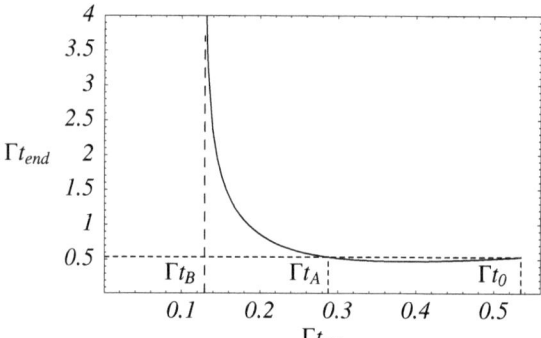

Figure 5.2: The time for the end of entanglement t_{end} is plotted against t_{sw} the time of switching states $|1\rangle$ and $|0\rangle$ in Eq. (5.1). Starting on the right at switching times of $t_0 \approx 0.5348/\Gamma$, the curve has a broad and small dip before rising rapidly to infinite time at $t_B \approx 0.1293/\Gamma$.

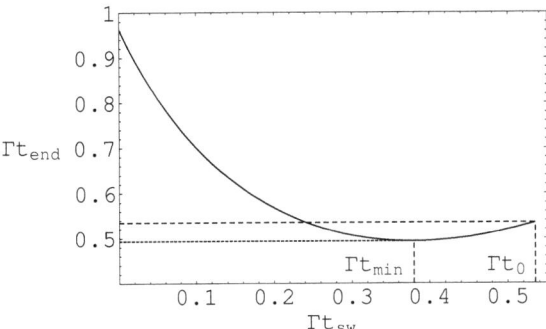

Figure 5.3: As in Figure 5.2 if the switch is done at only one end. Note that the maximum delay, which occurs at $t_{sw} = 0$, is now finite.

Alice or Bob makes the local unitary σ_x transformation, also to alter the end of entanglement. Now, the parameters a and c in Eq. (5.1) are interchanged, as also b and d, while z moves to the corners of the anti-diagonal. The roles of (a, d) and (b, c) in Eq. (5.3) are interchanged and we find that sudden death is hastened or delayed depending on the time of switching but it is now no longer averted indefinitely.

Figure 5.3 shows the results to be contrasted with those in Figure 5.2. A maximum, but still finite, delay is obtained for the earliest switch at $\Gamma t_{sw} = 0$, its value $\Gamma t_{end} = \ln(3 + \sqrt{5})/2 \approx 0.9624$ being a little less than double that of Γt_0. A simple, analytical expression describes the curve in Figure 5.3. With $x = \exp(-\Gamma t_{sw})$, $y = \exp(-\Gamma t_{end})$, we have

$$y(x) = \frac{3 - \sqrt{9 - 24x + 20x^2}}{2(2 - x)}. \tag{5.5}$$

The value of Γt_0 in Eq. (5.4) corresponds to the root $y = x = 2 - \sqrt{2}$.

In summary, we have shown that a simple local unitary operation that can be carried out on both qubits of an entangled pair changes the subsequent evolution of their entanglement. For mixed states under conditions which lead to sudden death of entanglement, such an operation can either hasten or delay sudden death, depending on the time at which it is carried out. There is a critical time before which the operation can even completely avert sudden death of entanglement. If the local transformation is done at only one of the qubits, sudden death is hastened or delayed but not averted completely.

5.2 Manipulating entanglement sudden death in zero- and finite-temperature reservoirs

Sudden death of entanglement is a serious limiting factor for the use of entangled qubits in quantum information processing. It would seem important to stabilize quantum systems against this unwanted phenomenon. In Section 5.1, we have addressed the practically relevant question whether it is possible to delay or even avert sudden death by application of particularly chosen local unitary transformations for a given initial state and a given open-system dynamics [102]. We have demonstrated that this is indeed possible for the special two-qubit system investigated first by Yu-Eberly [61] and have found similar effects in qubit-qutrit systems [137] (See Chapter 6).

In this section, we generalize our results of Section 5.1 and investigate finite-time disentanglement of two-qubits interacting with statistically independent (bosonic) reservoirs at finite temperatures. We demonstrate that based on the Peres-Horodecki criterion [37, 38] and on recent results of Huang and Zhu [90], it is possible to develop systematically a simple criterion capable of characterizing delay and avoidance of finite-time disentanglement of initially prepared two-qubit X-states in this open quantum system. With the help of this criterion we prove that, in agreement with recent conjectures based on numerical case studies [94], all initially prepared two-qubit X-states exhibit sudden death if at least one of the statistically independent reservoirs is at finite-temperature. However, if both reservoirs are at zero-temperature, there are always some X-states for which sudden death does not occur. Based on this criterion, we demonstrate that even at finite-temperatures of the reservoirs it is possible to hasten or to delay sudden death of entanglement. The characteristic time of sudden death can be controlled by the time when appropriate local unitary operations are applied. However, unlike in the zero-temperature case, when at least one of the reservoirs is at finite-temperature it is not possible to avoid sudden death completely by any choice of local unitary transformations.

5.2.1 Open-system dynamics of two-qubits coupled to statistically independent thermal reservoirs

In this section, we briefly summarize the basic equations of motion governing our open quantum system of interest. In order to put the problem of finite-time disentanglement, its delay, and avoidance into perspective, let us consider two non-interacting qubits which are spatially well separated so that each of them interacts with its own thermal reservoir (see Figure 5.4).

These two reservoirs are assumed to be statistically independent and are possibly also at different temperatures. The coupling of these two qubits

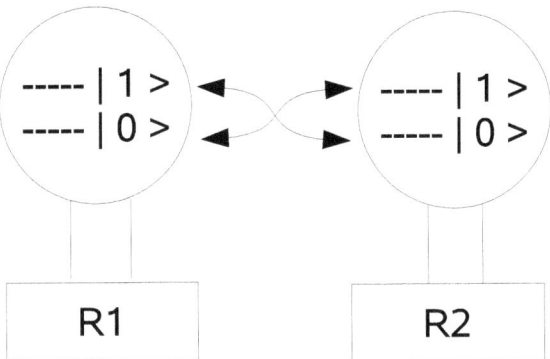

Figure 5.4: Schematic representation of two non-interacting two-level atoms (qubits) A and B initially prepared in an entangled state. Each of them interacts with its own local reservoir R_1 and R_2.

to the reservoirs can originate physically from the coupling of two two-level atoms to the (resonant) modes of the electromagnetic radiation field, for example, with the local radiation field at thermal equilibrium.

In the interaction picture and in the dipole- and rotating wave approximation, the resulting equation of motion of these two qubits is given by the master equation [57, 94]

$$\begin{aligned}\frac{d\rho}{dt} &= \frac{\gamma_1}{2}(m+1)[2\sigma_-^1 \rho \sigma_+^1 - \sigma_+^1 \sigma_-^1 \rho - \rho \sigma_+^1 \sigma_-^1] + \frac{\gamma_1}{2}m[2\sigma_+^1 \rho \sigma_-^1 - \sigma_-^1 \sigma_+^1 \rho \\ &\quad - \rho \sigma_-^1 \sigma_+^1] + \frac{\gamma_2}{2}(n+1)[2\sigma_-^2 \rho \sigma_+^2 - \sigma_+^2 \sigma_-^2 \rho - \rho \sigma_+^2 \sigma_-^2] \\ &\quad + \frac{\gamma_2}{2}n\,[2\sigma_+^2 \rho \sigma_-^2 - \sigma_-^2 \sigma_+^2 \rho - \rho \sigma_-^2 \sigma_+^2]. \end{aligned} \quad (5.6)$$

For reservoirs representing an electromagnetic radiation field in thermal equilibrium, m and n denote the mean photon numbers of the local reservoirs coupling to qubits 1 and 2. The spontaneous emission of atom i from its excited state $|1_i\rangle$ to its ground state $|0_i\rangle$ is described by the spontaneous decay rate γ_i and $\sigma_\pm^i$ are the corresponding raising and lowering operators, i.e., $\sigma_+^i = |1_i\rangle\langle 0_i|$ and $\sigma_-^i = |0_i\rangle\langle 1_i|$. The orthonormal atomic eigenstates $|1\rangle_{AB} = |1,1\rangle_{AB}$, $|2\rangle_{AB} = |1,0\rangle_{AB}$, $|3\rangle_{AB} = |0,1\rangle_{AB}$, $|4\rangle_{AB} = |0,0\rangle_{AB}$ form the (computational) basis of the four dimensional Hilbert space of the two qubits. The derivation of Eq. (5.6) also assumes the validity of the Born-Markov approximation. The most general solution of Eq. (5.6) for any initially prepared two-qubit quantum state ρ with density matrix elements ρ_{ij} in the computational basis is given as

$$\rho_{11}(t) = \frac{1}{(2m+1)(2n+1)}\{mn + m[(n+1)\rho_{11} + \rho_{33} - n(\rho_{22} - \rho_{33} + \rho_{44})]e^{-(2n+1)\gamma_2 t} + n[(m+1)\rho_{11} + (m+1)\rho_{22} - m(\rho_{33} + \rho_{44})] \times e^{-(2m+1)\gamma_1 t} + [(m+1)(n+1)\rho_{11} - m\rho_{33} - n(\rho_{22} + m\rho_{22} + m\rho_{33} - m\rho_{44})]e^{-[(2m+1)\gamma_1 + (2n+1)\gamma_2]t}\}, \quad (5.7)$$

$$\rho_{22}(t) = \frac{1}{(2m+1)(2n+1)}\{m(n+1) - m[(n+1)\rho_{11} + \rho_{33} - n(\rho_{22} - \rho_{33} + \rho_{44})]e^{-(2n+1)\gamma_2 t} + (n+1)[(m+1)\rho_{11} + (m+1)\rho_{22} - m(\rho_{33} + \rho_{44})]e^{-(2m+1)\gamma_1 t} + [-(m+1)(n+1)\rho_{11} + m\rho_{33} + n((m+1)\rho_{22} + m\rho_{33} - m\rho_{44})]e^{-[(2m+1)\gamma_1 + (2n+1)\gamma_2]t}\}, \quad (5.8)$$

$$\rho_{33}(t) = \frac{1}{(2m+1)(2n+1)}\{n(m+1) + (m+1)[(n+1)\rho_{11} + \rho_{33} - n(\rho_{22} - \rho_{33} + \rho_{44})]e^{-(2n+1)\gamma_2 t} - n[(m+1)\rho_{11} + (m+1)\rho_{22} - m(\rho_{33} + \rho_{44})]e^{-(2m+1)\gamma_1 t} + [-(m+1)(n+1)\rho_{11} + m\rho_{33} + n((m+1)\rho_{22} + m\rho_{33} - m\rho_{44})]e^{-[(2m+1)\gamma_1 + (2n+1)\gamma_2]t}\}, \quad (5.9)$$

$$\rho_{44}(t) = \frac{1}{(2m+1)(2n+1)}\{(m+1)(n+1) - (m+1)[(n+1)\rho_{11} + \rho_{33} - n(\rho_{22} - \rho_{33} + \rho_{44})]e^{-(2n+1)\gamma_2 t} - (n+1)[(m+1)\rho_{11} + (m+1)\rho_{22} - m(\rho_{33} + \rho_{44})]e^{-(2m+1)\gamma_1 t} + [(m+1)(n+1)\rho_{11} - m\rho_{33} - n((m+1)\rho_{22} + m\rho_{33} - m\rho_{44})]e^{-[(2m+1)\gamma_1 + (2n+1)\gamma_2]t}\}, \quad (5.10)$$

$$\rho_{12}(t) = \frac{1}{2m+1}\{m(\rho_{12} + \rho_{34})e^{\frac{-1}{2}(2n+1)\gamma_2 t} + [(m+1)\rho_{12} - m\rho_{34}] \times e^{\frac{-1}{2}[2(2m+1)\gamma_1 + (2n+1)\gamma_2]t}\}, \quad (5.11)$$

$$\rho_{13}(t) = \frac{1}{2n+1}\Big\{ n(\rho_{13}+\rho_{24})\,e^{\frac{-1}{2}(2m+1)\gamma_1 t} + [(n+1)\rho_{13} - n\rho_{24}]$$
$$\times e^{\frac{-1}{2}[(2m+1)\gamma_1 + 2(2n+1)\gamma_2]t} \Big\}, \tag{5.12}$$

$$\rho_{24}(t) = \frac{1}{2n+1}\Big\{ (n+1)(\rho_{13}+\rho_{24})\,e^{\frac{-1}{2}(2m+1)\gamma_1 t} + [n\rho_{24} - (n+1)\rho_{13}]$$
$$\times e^{\frac{-1}{2}[(2m+1)\gamma_1 + 2(2n+1)\gamma_2]t} \Big\}, \tag{5.13}$$

$$\rho_{34}(t) = \frac{1}{2m+1}\Big\{ (m+1)(\rho_{12}+\rho_{34})\,e^{\frac{-1}{2}(2n+1)\gamma_2 t} + [m\rho_{34} - (m+1)\rho_{12}]$$
$$\times e^{\frac{-1}{2}[2(2m+1)\gamma_1 + (2n+1)\gamma_2]t} \Big\}, \tag{5.14}$$

$$\rho_{14}(t) = \rho_{14}\,e^{-[(m+\frac{1}{2})\gamma_1 + (n+\frac{1}{2})\gamma_2]t}, \tag{5.15}$$

$$\rho_{23}(t) = \rho_{23}\,e^{-[(m+\frac{1}{2})\gamma_1 + (n+\frac{1}{2})\gamma_2]t}. \tag{5.16}$$

There is some numerical evidence [94] that the presence of non-zero mean thermal photon numbers in Eq. (5.6) may be responsible for sudden death. However, systematic exploration of these phenomena which is capable of proving the sufficiency of non-zero photon numbers for sudden death and of providing a systematic analytical understanding of sudden death in zero-temperature reservoirs has been missing so far. A major purpose of this section is to close this gap. In particular, we shall develop a simple analytical criterion for finite-time disentanglement which allows a systematic understanding of sudden death and its delay and avoidance for initially prepared entangled X-states.

5.2.2 The Peres-Horodecki criterion and entanglement sudden death

In this section, we analyze conditions under which initially prepared entangled two-qubit states evolving according to Eq. (5.6) exhibit entanglement sudden death. Starting from the general time dependent solution of these equations, sudden death is analyzed with the help of the Peres-Horodecki criterion [37, 38] and with the help of all principal minors of the partially transposed time-dependent two-qubit quantum state.

Recently, Huang and Zhu [90] studied the Peres-Horodecki criterion by focussing on the principal minors of the partially transposed density matrix. The principal minor $[\rho^{PT}(ijkl...)]$ of the partially transposed density operator ρ^{PT} is the determinant of the submatrix $\mathcal{M}(ijkl...)$ formed by the matrix elements of the $i,j,k,l,...$-th rows and columns of the partial transpose ρ^{PT}, i. e.,

$$\mathcal{M}(ijkl...) = \begin{pmatrix} \rho^{PT}_{ii}, & \rho^{PT}_{ij}, & \rho^{PT}_{ik}, & \rho^{PT}_{il}, & \cdots \\ \rho^{PT}_{ji}, & \rho^{PT}_{jj}, & \rho^{PT}_{jk}, & \rho^{PT}_{jl}, & \cdots \\ \rho^{PT}_{ki}, & \rho^{PT}_{kj}, & \rho^{PT}_{kk}, & \rho^{PT}_{jl}, & \cdots \\ \rho^{PT}_{li}, & \rho^{PT}_{lj}, & \rho^{PT}_{lk}, & \rho^{PT}_{ll}, & \cdots \\ \cdots & \cdots & \cdots & \cdots & \cdots \end{pmatrix}. \quad (5.17)$$

In general, if a matrix is positive semidefinite, then all its principal minors are non-negative, and vice versa [138]. Therefore, for an entangled two-qubit state ρ, the smallest principal minor of its partially transposed density operator must be negative. For two-qubit states non-negativity of the principal minors $[\rho^{PT}(1)]$, $[\rho^{PT}(2)]$, $[\rho^{PT}(3)]$, $[\rho^{PT}(4)]$, $[\rho^{PT}(12)]$, $[\rho^{PT}(13)]$, $[\rho^{PT}(24)]$, and $[\rho^{PT}(34)]$ is guaranteed already by non-negativity of the original density matrix ρ. As a consequence, a general two-qubit state is entangled if and only if the minimum value of the remaining seven principal minors $P(\rho^{PT})$ is negative, i. e.,

$$P(\rho^{PT}) \equiv \min\{ [\rho^{PT}(14)], [\rho^{PT}(23)], [\rho^{PT}(123)], [\rho^{PT}(124)], [\rho^{PT}(134)],$$
$$[\rho^{PT}(234)], [\rho^{PT}(1234)] \} < 0. \quad (5.18)$$

Typically, the investigation of these seven principal minors for general solutions $\rho(t)$ of the density operator equation (5.6) is cumbersome. However, significant simplifications are possible for identical zero-temperature reservoirs with $m = n = 0$, and $\gamma_1 = \gamma_2 = \gamma$ in Eq. (5.6). Thus, Huang and Zhu [90] could demonstrate that for asymptotically long times t_∞^- with $\gamma t_\infty^- \gg 1$, the separability of $\rho(t_\infty^-)$ is determined by the initial state ρ. In particular, these authors showed that the matrix

$$\tilde{\rho} = \begin{pmatrix} \rho_{11} & \rho_{21} & \rho_{13} & \rho_{23} \\ \rho_{12} & \rho_{11} + \rho_{22} & \rho_{14} & \rho_{13} + \rho_{24} \\ \rho_{31} & \rho_{41} & \rho_{11} + \rho_{33} & \rho_{21} + \rho_{43} \\ \rho_{32} & \rho_{31} + \rho_{42} & \rho_{12} + \rho_{34} & 1 \end{pmatrix}, \quad (5.19)$$

which is defined in terms of the matrix elements of the initially prepared two-qubit quantum state ρ, determines the asymptotic separability of the two qubits and thus the presence or absence of sudden death. The derivation of this result Eq. (5.19) is summarized in the following.

Derivation of Eq. (5.19)

For the simple case of vacuum reservoirs, i.e., $m = n = 0$, and $\gamma_1 = \gamma_2 = \gamma$ the time evolution of the two-qubit state $\tilde{\rho}(t)$ under amplitude damping can be written as [90]

$$\tilde{\rho}(t) = \begin{pmatrix} p^2 \rho_{11} & p\sqrt{p}\,\rho_{12} & p\sqrt{p}\,\rho_{13} & p\,\rho_{14} \\ p\sqrt{p}\,\rho_{21} & \rho_{22}(t) & p\,\rho_{23} & \rho_{24}(t) \\ p\sqrt{p}\,\rho_{31} & p\,\rho_{32} & \rho_{33}(t) & \rho_{34}(t) \\ p\,\rho_{41} & \rho_{42}(t) & \rho_{43}(t) & \rho_{44}(t) \end{pmatrix}, \tag{5.20}$$

where $p = e^{-\gamma t}$, and $\rho_{22}(t) = p(\,(1-p)\rho_{11}+\rho_{22}\,)$, $\rho_{24}(t) = \sqrt{p}(\,(1-p)\rho_{13}+\rho_{24}\,)$, $\rho_{33}(t) = p(\,(1-p)\rho_{11}+\rho_{33}\,)$, $\rho_{34}(t) = \sqrt{p}(\,(1-p)\rho_{12}+\rho_{34}\,)$, and $\rho_{44}(t) = 1 - p\,\rho_{22} - p\,\rho_{33} - (2p-p^2)\,\rho_{11}$. As $t \to \infty$, $p \to 0$ and the final state of the two-qubit system will always be in the ground $|0,0\rangle$ state, irrespective of its initial state.

The partial transpose of Eq. (5.20) is given by

$$\tilde{\rho}^{PT}(t) = \begin{pmatrix} p^2 \rho_{11} & p\sqrt{p}\,\rho_{21} & p\sqrt{p}\,\rho_{13} & p\,\rho_{23} \\ p\sqrt{p}\,\rho_{12} & \rho_{22}(t) & p\,\rho_{14} & \rho_{24}(t) \\ p\sqrt{p}\,\rho_{31} & p\,\rho_{41} & \rho_{33}(t) & \rho_{43}(t) \\ p\,\rho_{32} & \rho_{42}(t) & \rho_{34}(t) & \rho_{44}(t) \end{pmatrix}. \tag{5.21}$$

Due to exponential decay the probability p tends to zero for large times and the positivity or negativity of each principal minor of Eq. (5.21) is determined by its nonvanishing terms of lowest order in p. Based on this observation Eq. (5.21) can be written as

$$\tilde{\rho}^{PT}(t) = \bar{\rho} + \breve{\rho}, \tag{5.22}$$

with

$$\bar{\rho} = \begin{pmatrix} p^2 \rho_{11} & p\sqrt{p}\,\rho_{21} & p\sqrt{p}\,\rho_{13} & p\,\rho_{23} \\ p\sqrt{p}\,\rho_{12} & p(\rho_{11}+\rho_{22}) & p\,\rho_{14} & \sqrt{p}(\rho_{13}+\rho_{24}) \\ p\sqrt{p}\,\rho_{31} & p\,\rho_{41} & p(\rho_{11}+\rho_{33}) & \sqrt{p}(\rho_{21}+\rho_{43}) \\ p\,\rho_{32} & \sqrt{p}(\rho_{31}+\rho_{42}) & \sqrt{p}(\rho_{12}+\rho_{34}) & 1 \end{pmatrix}, \tag{5.23}$$

and

$$\breve{\rho} = \begin{pmatrix} 0 & 0 & 0 & 0 \\ 0 & -p^2\rho_{11} & 0 & -p\sqrt{p}\rho_{13} \\ 0 & 0 & -p^2\rho_{11} & -p\sqrt{p}\rho_{21} \\ 0 & -p\sqrt{p}\rho_{31} & -p\sqrt{p}\rho_{12} & -\breve{\rho}_{44} \end{pmatrix}, \tag{5.24}$$

and with $\check{\rho}_{44} = p(\rho_{11} + \rho_{22}) + p(\rho_{11} + \rho_{33}) - p^2\rho_{11}$. The matrix $\check{\rho}$ is negative semidefinite, i.e., $\check{\rho} \leq 0$.

Let us first of all assume that the minimum principal minor of the matrix (5.23) is nonzero, $P(\bar{\rho}) \neq 0$. Because the contributions of the matrix elements of $\check{\rho}$ are of higher order in $\sqrt{p}$ in comparison with $\bar{\rho}$, the minimum principal minor of the matrix (5.21) has the same sign as $P(\bar{\rho})$. Consequently, the matrix $\check{\rho}$ plays no role in determining the separability of $\tilde{\rho}(t)$. It is found that every principal minor of the matrix

$$\rho' = \begin{pmatrix} \rho_{11} & \rho_{21} & \rho_{13} & \rho_{23} \\ \rho_{12} & \rho_{11} + \rho_{22} & \rho_{14} & \rho_{13} + \rho_{24} \\ \rho_{31} & \rho_{41} & \rho_{11} + \rho_{33} & \rho_{21} + \rho_{43} \\ \rho_{32} & \rho_{31} + \rho_{42} & \rho_{12} + \rho_{34} & 1 \end{pmatrix}, \quad (5.25)$$

is equal to the corresponding principal minor of matrix (5.23) apart from multiplication by a suitable positive coefficient, such as

$$[\bar{\rho}(23)] = p^2\left[(\rho_{11} + \rho_{22})(\rho_{11} + \rho_{33}) - |\rho_{14}|^2\right] = p^2\left[\rho'(23)\right]. \quad (5.26)$$

Hence the separability of $\tilde{\rho}(t)$ is determined by the matrix ρ' which is independent of the decay parameter p.

If the minimum principal minor of the matrix (5.23) is zero, the higher order terms in matrix (5.24) must be taken into account and the corresponding principal minors of matrix (5.21) are either zero or nonpositive. For example, suppose $[\bar{\rho}(23)] = 0$, then the corresponding principal minor $[\tilde{\rho}^{PT}(23)]$ is given by

$$[\tilde{\rho}^{PT}(23)] = [\bar{\rho}(23)] - p^2\,\rho_{11}\,[\bar{\rho}(2)] - p^2\,\rho_{11}\,[\bar{\rho}(3)] + (p^2\,\rho_{11})^2. \quad (5.27)$$

For $\rho_{11} > 0$, $[\tilde{\rho}^{PT}(23)]$ is negative and for $\rho_{11} = 0$, $[\tilde{\rho}^{PT}(23)] = 0$. Similarly all seven principal minors are nonpositive if the corresponding principal minors of $\bar{\rho}$ are zero.

If $P(\tilde{\rho}) < 0$, the two-qubit state $\rho(t_\infty^-)$ is entangled and sudden death does not occur. If $P(\tilde{\rho}) > 0$, the asymptotic quantum state $\rho(t_\infty^-)$ is separable and sudden death takes place. In summary, Huang and Zhu [90] showed that for identical zero-temperature reservoirs the necessary and sufficient condition for sudden death is given by

$$P(\tilde{\rho}) > 0, \quad (5.28)$$

and is determined by the initially prepared two-qubit quantum state.

The derivation carried out to reach Eq. (5.28) is for the simplest case of vacuum reservoirs and by taking $\gamma_1 = \gamma_2 = \gamma$. For $\gamma_1 \neq \gamma_2$, it is not easy to arrive at the similar result (5.28). The situation becomes even more cumbersome for cases with $m \neq 0$ and $n \neq 0$, i.e., for thermal reservoirs. However, in the case of initially prepared X-states of the form of Eq. (5.1) the calculation of principal minors simplifies considerably as shown in the following. For initially prepared X-states, i.e., quantum states with $\rho_{12} = \rho_{13} = \rho_{24} = \rho_{34} = 0$ in the basis of Section 5.2.1, the condition (5.28) can be simplified considerably. In the subsequent subsection, all relevant seven principal minors for X-states ρ_X are evaluated.

Principal minors for X-states

For X-states the dependence of all seven principal-minors of Eq. (5.18) is given by

$$[\rho^{PT}(14)] = \begin{vmatrix} \rho_{11} & \rho_{23} \\ \rho_{32} & \rho_{44} \end{vmatrix} = \rho_{11}\rho_{44} - |\rho_{23}|^2,$$

$$[\rho^{PT}(23)] = \begin{vmatrix} \rho_{22} & \rho_{14} \\ \rho_{41} & \rho_{33} \end{vmatrix} = \rho_{22}\rho_{33} - |\rho_{14}|^2,$$

$$[\rho^{PT}(123)] = \begin{vmatrix} \rho_{11} & 0 & 0 \\ 0 & \rho_{22} & \rho_{14} \\ 0 & \rho_{41} & \rho_{33} \end{vmatrix} = \rho_{11}\,[\rho^{PT}(23)],$$

$$[\rho^{PT}(124)] = \begin{vmatrix} \rho_{11} & 0 & \rho_{23} \\ 0 & \rho_{22} & 0 \\ \rho_{32} & 0 & \rho_{44} \end{vmatrix} = \rho_{22}\,[\rho^{PT}(14)],$$

$$[\rho^{PT}(134)] = \begin{vmatrix} \rho_{11} & 0 & \rho_{23} \\ 0 & \rho_{33} & 0 \\ \rho_{32} & 0 & \rho_{44} \end{vmatrix} = \rho_{33}\,[\rho^{PT}(14)],$$

$$[\rho^{PT}(234)] = \begin{vmatrix} \rho_{22} & \rho_{14} & 0 \\ \rho_{41} & \rho_{33} & 0 \\ 0 & 0 & \rho_{44} \end{vmatrix} = \rho_{44}\,[\rho^{PT}(23)],$$

$$[\rho^{PT}(1234)] = \begin{vmatrix} \rho_{11} & 0 & 0 & \rho_{23} \\ 0 & \rho_{22} & \rho_{14} & 0 \\ 0 & \rho_{41} & \rho_{33} & 0 \\ \rho_{32} & 0 & 0 & \rho_{44} \end{vmatrix}$$
$$= [\rho^{PT}(14)] [\rho^{PT}(23)]. \tag{5.29}$$

From these expressions it is apparent that for initially prepared X-states, all the seven relevant principal minors are positive if and only if the two principal minors are positive, i.e.,

$$[\rho_X^{PT}(14)], \quad [\rho_X^{PT}(23)] > 0. \tag{5.30}$$

It is apparent from the general solution of the density operator of Eq. (5.6) as given by Eqs. (5.7-5.16) that an initially prepared two-qubit X-state remains an X-state for all times. This fact is valid for both vacuum and thermal reservoirs. However, for thermal reservoirs, the corresponding two principal minors can not be written in terms of matrix elements of the initial state alone but they also involve the quantities m, n, γ_1, and γ_2, in a non-trivial way.

Combining this observation with the results of Eq. (5.29) leads to the general conclusion that initially prepared two-qubit X-states exhibit sudden death if and only if at asymptotic times t_∞^- the principal minors are both positive, i.e.,

$$[\rho_X^{PT}(14)(t_\infty^-)] \quad , \quad [\rho_X^{PT}(23)(t_\infty^-)] > 0, \tag{5.31}$$

which are defined in terms of the matrix elements of the initially prepared two-qubit quantum state ρ. This condition determines the asymptotic separability of the two qubits and thus the presence or absence of sudden death.

5.2.3 Two-qubit X-states and quantum control of entanglement sudden death

In this section, we specialize our discussion of sudden death to initially prepared arbitrary two-qubit X-states. Delay and avoidance of sudden death of initially prepared two-qubit X-states coupled to two statistically independent zero-temperature reservoirs is discussed in Section 5.3. In Section 5.4 these results are generalized to reservoirs at finite temperatures. In particular, it is proved that if at least one of the reservoirs has non-zero temperature, all initially prepared X-states exhibit sudden death.

Let us first of all briefly summarize some basic properties of X-states. The density matrix of a two-qubit X-state is of the general form

$$\rho_X = \begin{pmatrix} \rho_{11} & 0 & 0 & \rho_{14} \\ 0 & \rho_{22} & \rho_{23} & 0 \\ 0 & \rho_{32} & \rho_{33} & 0 \\ \rho_{41} & 0 & 0 & \rho_{44} \end{pmatrix}, \qquad (5.32)$$

i.e., $\rho_{12} = \rho_{13} = \rho_{24} = \rho_{34} = 0$. These states are not unusual but arises naturally in a wide variety of physical situations [81, 139, 140]. In particular, Werner states [36] are special cases of such X-states and some aspects of their sudden death have been discussed already [62, 75, 80, 84]. Eq. (5.32) describes a quantum state provided the unit trace and positivity conditions $\sum_{i=1}^{4} \rho_{ii} = 1$, $\rho_{22}\rho_{33} \geq |\rho_{23}|^2$, and $\rho_{11}\rho_{44} \geq |\rho_{14}|^2$ are fulfilled. X-states are entangled if and only if either $\rho_{22}\rho_{33} < |\rho_{14}|^2$ or $\rho_{11}\rho_{44} < |\rho_{23}|^2$. Both conditions cannot hold simultaneously as there can be at most one negative eigenvalue for two-qubit states [136].

5.3 Delaying, hastening, and avoiding sudden death of entanglement in statistically independent vacuum reservoirs

As discussed in the previous section, for zero-temperature reservoirs the criterion for sudden death is given by Eq. (5.30) which together with the result of Eq. (5.25) yields the necessary and sufficient conditions for sudden death:

$$\begin{aligned}{}[\tilde{\rho}(14)] &= \rho_{11} - |\rho_{23}|^2 > 0, \\ [\tilde{\rho}(23)] &= (\rho_{11} + \rho_{22})(\rho_{11} + \rho_{33}) - |\rho_{14}|^2 > 0. \end{aligned} \qquad (5.33)$$

Depending on the degree of entanglement of the initially prepared two-qubit state, two different cases can be distinguished.

Case 1

For initially prepared entangled two-qubit states fulfilling the condition,

$$\rho_{11}\rho_{44} < |\rho_{23}|^2, \qquad (5.34)$$

the analytical expression for *negativity* of the quantum state $\rho(t)$ satisfying Eq. (5.6) for $\gamma_1 = \gamma_2 = \gamma$ and $m = n = 0$ is given by

$$N_1(\rho_X(t)) = \max\left\{0, \sqrt{F^2(p, \rho_{ii}) - 4p^2\left(\rho_{11}F(p, \rho_{ii}) - p^2\rho_{11}^2 - |\rho_{23}|^2\right)} - F(p, \rho_{ii})\right\}, \quad (5.35)$$

with $F(p, \rho_{ii}) = (1 - 2p + 2p^2)\rho_{11} + (1 - p)(\rho_{22} + \rho_{33}) + \rho_{44}$, $p = \exp(-\gamma t)$. For any initially entangled two-qubit state ρ, Eq. (5.34) implies $\rho_{22}\rho_{33} \geq |\rho_{14}|^2$ so that one of the conditions of Eq. (5.33) is satisfied. Thus, provided also the other condition, namely $\rho_{11} > |\rho_{23}|^2$, is satisfied, an initially prepared entangled two-qubit state exhibits sudden death and its *negativity* becomes zero at a finite time, say t_1.

Provided both conditions of Eq. (5.33) are fulfilled, sudden death of entanglement can be delayed or even avoided by local unitary operations acting on the two qubits involved. In particular, let us concentrate on local unitary operations which exchange the density matrix elements $\rho_{11}(t)$ and $\rho_{44}(t)$ of the quantum state at a time $t < t_1$ in such a way that their product, i.e., $\rho_{11}(t)\rho_{44}(t)$, remains constant but that the condition $\rho_{44}(t) > |\rho_{23}(t)|^2$ is violated. According to Eq. (5.33), in such a case sudden death will be avoided. The matrix element $\rho_{44}(t)$ is the probability of finding both qubits in their ground states. Thus, as a consequence of the dynamics of Eq. (5.6), the density matrix element $\rho_{44}(t)$ increases monotonically. There will be a limiting time t_{sw} for any possible switching of these matrix elements for which sudden death can still be avoided. If the local operation is applied after this limiting time, sudden death may possibly be delayed but it is unavoidable.

This simple consequence of the criterion of Eq. (5.33) explains recent numerical work on this problem [102] (see Section 5.1). In fact, operations of this type can avoid sudden death for any initially prepared two-qubit X-state provided they are applied at a time t such that $t < t_{sw}$, where t_{sw} is the time of switching states $|0\rangle$ and $|1\rangle$. In particular, this applies to the subset of Werner states with $\rho_{14} = 0$ which are mixtures of a singlet state with probability a and a completely unpolarized (chaotic) state. These Werner states exhibit sudden death in the parameter range $a \in [1/3, (-1 + \sqrt{5})/2)$ [83] where $\rho_{11}(t) > |\rho_{23}(t)|^2$ while entanglement decays asymptotically for values of a in the range $a \in ((-1 + \sqrt{5})/2, 1]$ which corresponds to $\rho_{11}(t) < |\rho_{23}(t)|^2$.

Let us now deal with the question of which unitary transformations can achieve such a switch between $\rho_{44}(t)$ and $\rho_{11}(t)$. The most general 2×2

unitary matrix acting on a qubit is given by

$$U(2) = \begin{pmatrix} \cos(\theta)e^{i\alpha} & -\sin(\theta)e^{i(\alpha-\omega)} \\ \sin(\theta)e^{i(\beta+\omega)} & \cos(\theta)e^{i\beta} \end{pmatrix}, \quad (5.36)$$

which is a linear superposition of the Pauli matrices σ_x, σ_y, σ_z, and the Identity matrix σ_0. Exchanging the matrix elements ρ_{11} and ρ_{44} can be achieved by applying two appropriately chosen unitaries U_A and U_B of the form of Eq. (5.36) on qubits A and B at a suitably chosen time, say t, so that the X-state $\rho_X(t)$ is transformed into another X-state $(U_A \otimes U_B)\rho_X(t)(U_A^\dagger \otimes U_B^\dagger)$, for example.

The most general local unitary operations transforming an arbitrary X-state into another one fulfill the conditions

$$\sin(2\theta_A) = \sin(2\theta_B) = 0 \quad \longrightarrow \quad \theta_A = r_A\pi/2,\ \theta_B = r_B\pi/2, \quad (5.37)$$

with $r_A, r_B \in \mathbb{Z}$. X-state preserving local unitary transformations with even values of r_A and r_B do not have any significant effect on the density matrix elements except multiplying $\rho_{14}(t)$ by a constant phase factor. Odd values of r_A and r_B serve the purpose of exchanging $\rho_{11}(t)$ and $\rho_{44}(t)$. For any odd value of $r_A = r_B$, for example, the corresponding unitary two-qubit operator is given by

$$U = \begin{pmatrix} 0 & 0 & 0 & e^{2i(\alpha-\omega)} \\ 0 & 0 & -e^{i(\alpha+\beta)} & 0 \\ 0 & -e^{i(\alpha+\beta)} & 0 & 0 \\ e^{2i(\beta+\omega)} & 0 & 0 & 0 \end{pmatrix}. \quad (5.38)$$

A case in which such a X-state-preserving local unitary transformation is applied only onto qubit B can be described by parameters $\theta_A = \alpha_A = \beta_A = \omega_A = 0$, for example. They lead to the transformations $\rho_{11}(t) \Leftrightarrow \rho_{22}(t)$, $\rho_{33}(t) \Leftrightarrow \rho_{44}(t)$, and $\rho_{14}(t) \Leftrightarrow \rho_{23}(t)$. In view of the characteristic time evolution of $\rho_{22}(t)$ in zero-temperature reservoirs and the criterion of Eq. (5.33), this implies that such a switch of matrix elements may delay sudden death but it cannot be avoided.

Case 2

For initially prepared entangled two-qubit X-states satisfying the alternative condition

$$\rho_{22}\rho_{33} < |\rho_{14}|^2, \quad (5.39)$$

the expression for *negativity* of the resulting quantum state $\rho_X(t)$ is given by

$$N_2(\rho_X(t)) = \max\left\{0,\, p\left(\sqrt{(\rho_{22}-\rho_{33})^2 + 4|\rho_{14}|^2} - (\rho_{22}+\rho_{33})\right)\right.$$
$$\left. -(2-2p)\rho_{11}\right\}. \tag{5.40}$$

Eq. (5.39) implies $\rho_{11}\rho_{44} \geq |\rho_{23}|^2$. Thus, the first condition of Eq. (5.33) is always satisfied so that finite-time disentanglement occurs whenever also the second condition is satisfied. The simplest case arises for $\rho_{23} = 0$ where the initially prepared state is a Werner state, i.e., an incoherent mixture of a triplet state with probability a and the completely unpolarized state. Sudden death takes place in the parameter regime $a \in [1/3, 1)$ where both conditions are satisfied during the time evolution. In the case of an initially prepared Bell state, i.e., for $a = 1$, the second condition of Eq. (5.33) fails and entanglement decays asymptotically.

As discussed above, the first condition of Eq. (5.33) is always fulfilled in the cases considered here so that sudden death takes place always except in the particular case of an initially prepared Bell state which fulfills the condition $(\rho_{11}(t)+\rho_{22}(t))(\rho_{11}(t)+\rho_{33}(t)) = |\rho_{14}(t)|^2$. As a consequence any switch capable of exchanging $\rho_{23}(t)$ and $\rho_{14}(t)$ will be sufficient to avoid or delay sudden death. Such a switch can be implemented by a local unitary X-state-preserving transformation acting on qubit A or B only. As a result the second condition of Eq. (5.33) remains always true, while the validity of the first condition depends on the choice of the switching time t. If $\rho_{33}(t) > |\rho_{14}(t)|^2$ sudden death is unavoidable. However, for all switching times violating this condition sudden death is averted completely. As an example, let us consider Werner states with $\rho_{23}(0) = 0$. In the parameter range $a \in [1/3, (-1+\sqrt{5})/2)$ the condition $\rho_{33}(t) > |\rho_{14}(t)|^2$ is fulfilled for all times so that ESD takes place. In the parameter regime $a \in ((-1+\sqrt{5})/2, 1]$ this condition is violated so that entanglement decays asymptotically.

5.4 Hastening and delaying sudden death in statistically independent thermal reservoirs

According to Eq. (5.30), sudden death takes place if and only if the two principal minors

$$[\rho^{PT}(14)] = m^2(m+1)^2 + e^{-(2m+1)\gamma t}[F^{(1)}] + e^{-2(2m+1)\gamma t}[F^{(2)}]$$
$$+ e^{-3(2m+1)\gamma t}[F^{(3)}] + e^{-4(2m+1)\gamma t}[F^{(4)}], \tag{5.41}$$

and
$$\begin{aligned}[\rho^{PT}(23)] &= m^2(m+1)^2 + e^{-(2m+1)\gamma t}[G^{(1)}] + e^{-2(2m+1)\gamma t}[G^{(2)}]\\ &+ e^{-3(2m+1)\gamma t}[G^{(3)}] + e^{-4(2m+1)\gamma t}[G^{(4)}],\end{aligned} \quad (5.42)$$

are positive in the limit of very long interaction times. For simplicity, we have taken $m = n$ and $\gamma_1 = \gamma_2 = \gamma$ in Eq. (5.41) and Eq. (5.42). The quantities $F^{(i)}$ and $G^{(i)}$ are functions of m and of the initial matrix elements of the initially prepared quantum state. Their explicit forms are given by

$$F^{(1)} = m(m+1)\left[(2m+1)\rho_{11} + \rho_{22} + \rho_{33} - 2m\rho_{44}\right],$$

$$\begin{aligned}F^{(2)} &= -2m^4\left[2\rho_{44}^2 - \rho_{44} + \rho_{22} + 8|\rho_{23}|^2 + \rho_{33}\right] + 2m^3\left[-2\rho_{44}^2 + 2\rho_{33}\rho_{44}\right.\\ &\left.+ \rho_{44} - 16|\rho_{23}|^2 - 2\rho_{33} + 2\rho_{22}(\rho_{44} - 1)\right] - m^2\left[\rho_{22}^2 + (2\rho_{33}\right.\\ &\left.- 4\rho_{44} + 3)\rho_{22} + \rho_{33}^2 + 24|\rho_{23}|^2 + 3\rho_{33} - 4\rho_{33}\rho_{44} - \rho_{44}\right]\\ &- 4m(m+1)^3\rho_{11}^2 - m\left[\rho_{22}^2 + 2\rho_{33}\rho_{22} + \rho_{22} + \rho_{33}^2 + 8|\rho_{23}|^2\right.\\ &\left.+ \rho_{33}\right] - |\rho_{23}|^2 + (m+1)^2\rho_{11}\left[(8\rho_{44} + 2)m^2 +\right.\\ &\left.(-4\rho_{22} - 4\rho_{33} + 2)m + 1\right],\end{aligned}$$

$$\begin{aligned}F^{(3)} &= -2\rho_{11}^2(m+1)^3 + (m+1)\rho_{11}\left[(2m^2 + m - 1)(\rho_{22} + \rho_{33}) + 2m\rho_{44}\right]\\ &+ m\left[(m+1)\rho_{22}^2 + (2(m+1)\rho_{33} - m(2m+3)\rho_{44})\rho_{22} +\right.\\ &\left.(m+1)\rho_{33}^2 + 2m^2\rho_{44}^2 - m(2m+3)\rho_{33}\rho_{44}\right],\end{aligned}$$

$$F^{(4)} = \left[m^2(\rho_{11} - \rho_{22} - \rho_{33} + \rho_{44}) + m(2\rho_{11} - \rho_{22} - \rho_{33}) + \rho_{11}\right]^2. \quad (5.43)$$

Similarly, the expressions for $G^{(i)}$ in Eq. (5.42) are given by

$$G^{(1)} = m(m+1)\left[(2m+1)\rho_{11} + \rho_{22} + \rho_{33} - 2m\rho_{44}\right],$$

$$\begin{aligned}G^{(2)} &= -2m^4\left[2\rho_{22}^2 - 4\rho_{33}\rho_{22} - \rho_{22} + 2\rho_{33}^2 + \rho_{11} - \rho_{33} + 8|\rho_{14}|^2 + \rho_{44}\right]\\ &- 2m^3\left[4\rho_{22}^2 - 8\rho_{33}\rho_{22} - 2\rho_{22} + 4\rho_{33}^2 + 3\rho_{11} - 2\rho_{33} + 16|\rho_{14}|^2\right.\\ &\left.+ \rho_{44}\right] + m^2\left[\rho_{11}^2 - 2(\rho_{44} + 3)\rho_{11} - 6\rho_{22}^2 - 6\rho_{33}^2 + \rho_{44}^2 + 2\rho_{22}\right.\\ &\left.+ 12\rho_{22}\rho_{33} + 2\rho_{33} - 24|\rho_{14}|^2\right] + m\left[2\rho_{11}^2 + (\rho_{22} + \rho_{33} - 2\rho_{44}\right.\\ &\left.- 2)\rho_{11} - 2\rho_{22}^2 - 2\rho_{33}^2 + 4\rho_{22}\rho_{33} - 8|\rho_{14}|^2 - \rho_{22}\rho_{44} - \rho_{33}\rho_{44}\right]\\ &+ \rho_{11}^2 + \rho_{22}\rho_{33} + \rho_{11}(\rho_{22} + \rho_{33}) - |\rho_{14}|^2,\end{aligned}$$

$$\begin{aligned}G^{(3)} &= -2\rho_{11}^2(m+1)^3 + (m+1)\rho_{11}\left[(2m^2 + m - 1)(\rho_{22} + \rho_{33}) +\right.\\ &\left.2m\rho_{44}\right] + m\left[(m+1)\rho_{22}^2 + (2(m+1)\rho_{33} - m(2m+3)\rho_{44})\rho_{22}\right.\\ &\left.+ (m+1)\rho_{33}^2 + 2m^2\rho_{44}^2 - m(2m+3)\rho_{33}\rho_{44}\right],\end{aligned}$$

$$G^{(4)} = [m^2(\rho_{11} - \rho_{22} - \rho_{33} + \rho_{44}) + m(2\rho_{11} - \rho_{22} - \rho_{33}) + \rho_{11}]^2. \quad (5.44)$$

For sufficiently long times, say $t \geq t_\infty^-$ and for $m > 0$, factors of the form $e^{-(2m+1)\gamma t_\infty^-}$ are exponentially small and therefore both $[\rho^{PT}(14)]$ and $[\rho^{PT}(23)]$ are positive. Analogously, one can show that for unequal values of the mean photon numbers m and n, both minors are positive if and only if at least one of these mean photon numbers is not equal to zero. Hence, we arrive at the central result that if one of the (photon) reservoirs is at nonzero temperature all initially prepared X-states exhibit sudden death.

As sudden death is unavoidable in these cases, it may be useful at least to delay it. Indeed this can be achieved for all possible X-states [61, 102]. For the sake of demonstration let us consider the particular example of an initially prepared entangled state of the form

$$\rho = \frac{1}{3}(|1,1\rangle\langle 1,1| + 2|\Psi\rangle\langle\Psi|), \quad (5.45)$$

with $|\Psi\rangle = (|0,1\rangle \pm |1,0\rangle)/\sqrt{2}$. It is known that this entangled mixed state, while interacting with statistically independent vacuum reservoirs, looses its entanglement at $t \approx 0.5348/\gamma$ [102]. However, while interacting with independent reservoirs at finite temperatures, the time of sudden death for this initial state depends on the values of m and n. The solution of Eq. (5.6) for the input state of Eq. (5.45) can be obtained easily using the general solution. After taking the partial transpose of the resulting density operator it is possible to obtain an analytical expression for *negativity* of the quantum state at any time t. Setting $m = n = 0.1$, for example, we observe that sudden death occurs at time $t_{end} \approx 0.4115/\gamma$. Depending on the time when local unitary transformations are applied to qubits A and B, sudden death can be speeded up or delayed for some finite time.

Figure 5.5 displays the time t_{end} at which sudden death takes place and its dependence on the time of switching t_{sw}. The earlier appropriate local unitary transformations are applied, the more sudden death is delayed. However, typically such a delay is possible only for a certain range of switching times t_{sw}, such as $t_{sw} < t_B = 0.279/\gamma$ in Figure 5.5. Eventually sudden death is unavoidable. In the case considered in Figure 5.5, it takes place at $t_{end} \approx 0.9817/\gamma$.

In Figure 5.6 the relation between t_{end} and t_{sw} is depicted for mean thermal photon numbers $m = n = 0.01$. In this case sudden death occurs at $t_A \approx 0.5172/\gamma$. If the switch is applied before $t_A \approx 0.5172/\gamma$, sudden death is hastened. Any switch made before $t_B \approx 0.2877/\gamma$ delays sudden death up to the maximum possible time $t_{end} \approx 2.7087/\gamma$. This larger delay in comparison with the case considered in Figure 5.5 is due to smaller values of the mean photon numbers.

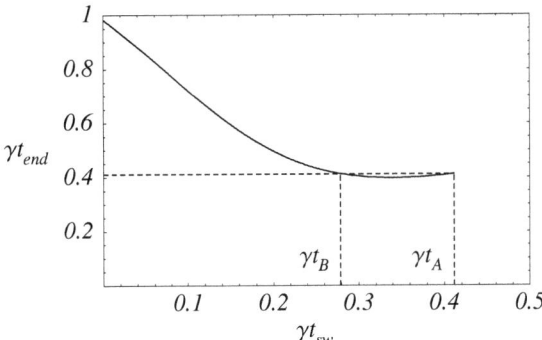

Figure 5.5: Dependence of the time of sudden death t_{end} on the switching time t_{sw} for mean photon numbers $m = n = 0.1$ of the statistically independent reservoirs: X-state-preserving local unitary transformations switch the density matrix elements ρ_{11} and ρ_{44} in Eq. (5.45). Around switching time $t_A \approx 0.4115/\gamma$ this dependence exhibits a broad and small dip before rising to the maximum possible time $t_{end} \approx 0.9817/\gamma$.

In summary we have presented a criterion characterizing the conditions which lead to entanglement sudden death of X-states of two qubits coupled to statistically independent reservoirs at finite temperatures. Based on this criterion, we have presented an analytical description of sudden death of X-states and its delaying or its avoidance by local unitary actions. We have proved that if at least one of the reservoirs is at finite temperature, all X-states exhibit sudden death. Thus, in these cases sudden death can only be delayed but not averted.

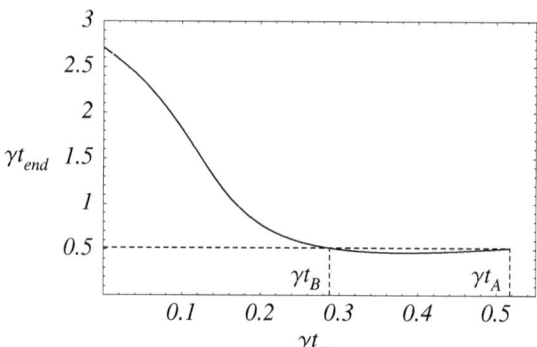

Figure 5.6: Dependence of the time of sudden death t_{end} on the switching time t_{sw} for mean photon numbers $m = n = 0.01$: The local unitary transformations are the same as in Figure 5.5. For switching times below $t_A \approx 0.5172/\gamma$ this dependence has a broad and small dip before rising to the maximum possible time $t_{end} \approx 2.7087/\gamma$.

Chapter 6

Manipulating entanglement sudden death of qubit-qutrit systems

In Chapter 5, we have demonstrated that we can indeed hasten, delay, or even completely avoid sudden death of two qubits interacting with statistically independent reservoirs. In this Chapter, we slightly enlarge the dimension of the Hilbert space to six, i.e., qubit (two dimensional quantum system) times qutrit (three dimensional quantum system). We first demonstrate the existence of sudden death in these systems due to amplitude damping and phase damping. We then discuss the possibility of accelerating, delaying or avoiding sudden death of entanglement by local unitary transformations.

This Chapter is organized as follows. In Section 6.1, we study the pure entangled states of qubit-qutrit systems and investigate the existence of sudden death of different initially entangled pure and mixed states interacting with statistically independent reservoirs at zero-temperature. In Section 6.4, we discuss sudden death due to phase damping. Hastening, delaying or completely avoiding sudden death is provided in Section 6.2. We discuss the resulting asymptotic states in Section 6.3.

6.1 Entanglement sudden death of qubit-qutrit systems by amplitude damping

So far two investigations on sudden death of entanglement of qubit-qutrit systems exist in the literature. Our investigation of this effect is for amplitude damping [92], while the second investigation deals with phase damping [100]. We first discuss the results for amplitude damping and later briefly comment on the work for phase damping.

6.1.1 Maximally entangled pure states for $2 \otimes 3$ systems

As discussed before, *negativity* [54] is a computable measure of entanglement and it completely describes all entangled states in the Hilbert space of qubit-qutrit systems. The Peres-Horodecki criterion tells us that for $2 \otimes 2$ and $2 \otimes 3$ Hilbert spaces all states having positive partial transpose (PPT) (therefore, zero *negativity*) are separable [37, 38].

Maximally entangled pure states in $2 \otimes 3$ Hilbert space are extensions of the Bell states. To describe the set of pure entangled states in this Hilbert space, an arbitrary pure state can be represented in terms of its Schmidt decomposition [29]

$$|\Psi\rangle = \sum_{ij} a_{ij} |i,j\rangle, \qquad (6.1)$$

where $i = 0, 1$ denotes the qubit indices and $j = 0, 1, 2$ represent the qutrit indices. For $|\Psi\rangle$ to be a valid state vector we need $\sum_{ij} |a_{ij}|^2 = 1$, where the amplitudes a_{ij} are complex numbers in general. The density operator ρ is defined by $\rho = |\Psi\rangle\langle\Psi|$. We take the partial transpose of ρ with respect to the qutrit basis. (Taking partial transpose with respect to the qubit basis gives the same result.) The spectrum of the partially transposed matrix ρ^{T_B} is given by

$$\left\{ 0,\, 0,\, -\sqrt{f(a_x)},\, \sqrt{f(a_x)},\, \frac{1}{2}(1 - \sqrt{1 - 4f(a_x)}),\, \frac{1}{2}(1 + \sqrt{1 - 4f(a_x)}) \right\}, \qquad (6.2)$$

where $a_x = a_{ij}$ corresponds to the six combinations i.e., $a_1 = a_{00}$, $a_2 = a_{01}$, $a_3 = a_{02}$, $a_4 = a_{10}$, $a_5 = a_{11}$, and $a_6 = a_{12}$.

Surprisingly, all the eigenvalues depend on the single function $f(a_x)$. The expression for $f(a_x)$ is given by

$$\begin{aligned} f(a_x) &= |a_2|^2 |a_4|^2 + |a_3|^2 |a_4|^2 - 2\mathrm{Re}(a_1 \bar{a}_2 \bar{a}_4 a_5) - 2\mathrm{Re}(a_1 \bar{a}_3 \bar{a}_4 a_6) + |a_1|^2 |a_5|^2 \\ &\quad + |a_3|^2 |a_5|^2 + |a_1|^2 |a_6|^2 + |a_2|^2 |a_6|^2 - 2\mathrm{Re}(a_2 \bar{a}_3 \bar{a}_5 a_6), \end{aligned} \qquad (6.3)$$

where $\bar{a}_i$ are the complex conjugates of a_i. As the partially transposed matrix of a Hermitian operator is also Hermitian, all its eigenvalues must be real. This implies that $0 \leq f(a_x) \leq 1/4$. At the lower bound, the pure states are separable, while at the upper bound, the pure states are maximally entangled. For $0 < f(a_x) < 1/4$, the pure states are non-maximally entangled. For a pure state to be entangled, at least two coefficients in Eq. (6.1) must be non-zero.

It is evident from the above discussion that the partial transpose of pure states can have only one negative eigenvalue. Therefore *negativity* for pure entangled states can be written as

$$N = 2\sqrt{f(a_x)}. \tag{6.4}$$

As *negativity* is invariant under all local unitary transformations, the square root of the function $f(a_x)$ is also invariant under all local unitary transformations. By the Schmidt decomposition of Eq. (6.1), an arbitrary qubit-qutrit state may be written in the form

$$|\Phi_1\rangle = (U_A \otimes U_B)\left(\alpha|0,0\rangle + \sqrt{1-\alpha^2}|1,1\rangle\right), \tag{6.5}$$

where U_A and U_B denote the unitary transformation from the local computational bases to the Schmidt basis of the qubit and the qutrit, respectively, and where $\alpha \in [0; 1/\sqrt{2}]$ and $|0,0\rangle = |0\rangle_A \otimes |0\rangle_B$, etc. Since f is invariant with respect to local unitary operations, we may ignore U_A and U_B for now. Using $a_1 = \alpha$ and $a_5 = \sqrt{1-\alpha^2}$, we find $f(a_x) = \alpha^2(1-\alpha^2)$, which attains its maximum value $1/4$ for $\alpha = 1/\sqrt{2}$. This shows that a qubit-qutrit state is maximally entangled with respect to *negativity*, if and only if it is of the form of Eq. (6.5) with $\alpha = 1/\sqrt{2}$.

Let us consider a specific pure state of a given degree of entanglement,

$$|\Phi_1\rangle = \alpha|0,0\rangle + \beta|1,1\rangle, \tag{6.6}$$

where $|\alpha|^2 + |\beta|^2 = 1$. Negativity of this state is given by $2\alpha\beta$. This value will remain invariant for all pure entangled states obtained by applying local unitaries to Eq. (6.6). Therefore, we can characterize the set of pure entangled states for a given degree of entanglement by Eq. (6.5). Some other examples of such states are

$$|\Phi_2^\pm\rangle = \alpha|0,1\rangle \pm \beta|1,2\rangle, \tag{6.7}$$

$$|\Phi_3^\pm\rangle = \alpha|0,2\rangle \pm \beta|1,0\rangle, \tag{6.8}$$

6.1.2 Three-level atom and quantum interference

Let us consider a three-level atom in the V configuration. Let $|e\rangle$, $|u\rangle$ be two nondegenerate excited states of the atom with transition frequencies to the ground state $|g\rangle$ given by ω_e, ω_u and the electric dipole moments $\vec{\mu}_e$, $\vec{\mu}_u$

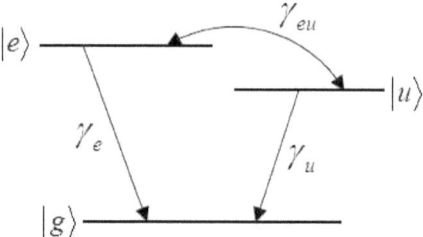

Figure 6.1: Three-level atom in the V configuration (system I).

respectively. We assume that the excited levels decay to the ground level $|g\rangle$ by spontaneous emission and a direct transition between the excited levels is forbidden. Spontaneous decay constants of the excited levels are given by [87]

$$\gamma_i = \frac{2|\vec{\mu}_i|^2 \omega_i^3}{3\hbar c^3}, \quad i = e, u. \tag{6.9}$$

If the dipole moments of the allowed transitions are parallel, then indirect coupling between levels $|e\rangle$ and $|u\rangle$ can appear due to interaction with the vacuum [86], i.e., quantum interference between transitions $|e\rangle \to |g\rangle$ and $|u\rangle \to |g\rangle$ with a cross damping constant

$$\gamma_{eu} = \frac{2\sqrt{\omega_e^3 \omega_u^3}}{3\hbar c^3} \vec{\mu}_e \cdot \vec{\mu}_u = \beta_I \sqrt{\gamma_e \gamma_u}, \tag{6.10}$$

where

$$\beta_I = \frac{\vec{\mu}_e \cdot \vec{\mu}_u}{|\vec{\mu}_e||\vec{\mu}_u|} = \cos\theta. \tag{6.11}$$

The parameter β_I represents the mutual orientation of the atomic transition dipole moments. If the dipole moments are parallel, $\beta_I = 1$, then the cross damping term (quantum interference) is maximal with $\gamma_{eu} = \sqrt{\gamma_e \gamma_u}$. If the dipole moments are perpendicular, $\beta_I = 0$, then $\gamma_{eu} = 0$, quantum interference vanishes (compare with Figure 6.1).

The master equation [141] for time evolution of Figure 6.1 is given by

$$\frac{d\rho}{dt} = -\mathrm{i}[H, \rho] + \Lambda_I \rho, \tag{6.12}$$

where the damping term is

$$\Lambda_I \rho = \frac{\gamma_e}{2}(2\sigma_{ge}\rho\sigma_{eg} - \sigma_{ee}\rho - \rho\sigma_{ee}) + \frac{\gamma_u}{2}(2\sigma_{gu}\rho\sigma_{ug} - \sigma_{uu}\rho - \rho\sigma_{uu}) + \frac{\gamma_{eu}}{2}$$
$$\times (2\sigma_{ge}\rho\sigma_{ug} - \sigma_{ue}\rho - \rho\sigma_{ue}) + \frac{\gamma_{eu}}{2}(2\sigma_{gu}\rho\sigma_{eg} - \sigma_{eu}\rho - \rho\sigma_{eu}). \quad (6.13)$$

Here, σ_{jk} is the atomic transition operator from the level $|k\rangle$ to the level $|j\rangle$.

In atomic spectroscopy, the atomic transition dipole moments are usually perpendicular, therefore the system described by Eq. (6.13) is difficult to realize. It was shown [141] that quantum interference can be duplicated to a large degree by three-level system II (see Figure 6.2) with the excited states $|2\rangle$, $|1\rangle$ and the ground state $|0\rangle$. The master equation for time evolution of this system contains damping operator

$$\Lambda_{II}\rho = \frac{\gamma_2}{2}(2\sigma_{02}\rho\sigma_{20} - \sigma_{22}\rho - \rho\sigma_{22}) + \frac{\gamma_1}{2}(2\sigma_{01}\rho\sigma_{10} - \sigma_{11}\rho - \rho\sigma_{11}). \quad (6.14)$$

To prove it, let us consider the system I with $\gamma_e = \gamma_u = \gamma$. We introduce symmetric and antisymmetric superposition of the excited states

$$|s\rangle = \frac{1}{\sqrt{2}}(|e\rangle + |u\rangle),$$
$$|a\rangle = \frac{1}{\sqrt{2}}(|e\rangle - |u\rangle), \quad (6.15)$$

and the ground state g as a new basis in $\mathbb{C}^3$. Eq. (6.13) can be written in this basis by

$$\tilde{\Lambda}_I \rho = \frac{\gamma}{2}(1+\beta_I)(\sigma_{gs}\rho\sigma_{sg} - \sigma_{ss}\rho - \rho\sigma_{ss}) + \frac{\gamma}{2}(1-\beta_I)$$
$$\times (\sigma_{ga}\rho\sigma_{ag} - \sigma_{aa}\rho - \rho\sigma_{aa}). \quad (6.16)$$

We note that for almost parallel dipole moments, $\beta_I \simeq 1$, antisymmetric level is metastable. So in this system maximum interference corresponds to the stability of level $|a\rangle$. The system II with damping operator Eq. (6.14), is a generalization of this system with the excited states $|2\rangle$ and $|1\rangle$ to be nondegenerate and their decay rates to the ground state $|0\rangle$ to be arbitrary. Moreover, the dipole moments can be perpendicular.

Quantum interference is defined by

$$\beta_{II} = \frac{\gamma_2 - \gamma_1}{\gamma_2 + \gamma_1} = \frac{1-k}{1+k}, \quad (6.17)$$

where $k = \gamma_1/\gamma_2$. Quantum interference appears for $\gamma_1 \ll \gamma_2$ provided $\Delta \ll \gamma_1, \gamma_2$, where Δ is the detuning between the two excited levels. So when the level $|1\rangle$ is metastable we expect maximal effects of quantum interference.

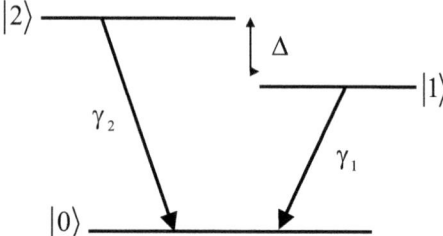

Figure 6.2: Three-level atom in the V configuration (system II).

6.1.3 Physical Model

We consider two atoms initially prepared in an entangled state. Both atoms interact with their own statistically independent vacuum reservoirs and they are separated in space by a large distance compared to the wavelength of the emitted radiation. This condition implies that both atoms no longer interact with each other. Our qubit is a two-level atom interacting with reservoir 1. Our qutrit is a three-level atom interacting with reservoir 2. We consider the three-level atom in the V configuration. Similar analysis has been done for qutrit-qutrit systems (two three-level atoms) [87]. Our principal system consists of two entangled atoms while both reservoirs serve as environments. Spontaneous emission of the excited states causes decoherence and the degradation of entanglement in the principal system. We are interested in studying the dynamics of this entanglement.

The master equation for the two separated atoms is given by

$$\frac{d\rho}{dt} = \Upsilon^{AB}\rho, \qquad (6.18)$$

where

$$\Upsilon^{AB}\rho = \frac{1}{2}\gamma(2\sigma^A_{01}\rho\sigma^A_{10} - \sigma^A_{11}\rho - \rho\sigma^A_{11}) + \frac{1}{2}\gamma_2(2\sigma^B_{02}\rho\sigma^B_{20} - \sigma^B_{22}\rho - \rho\sigma^B_{22}) + \frac{1}{2}\gamma_1(2\sigma^B_{01}\rho\sigma^B_{10} - \sigma^B_{11}\rho - \rho\sigma^B_{11}). \qquad (6.19)$$

Here

$$\sigma^A_{kl} = \sigma_{kl} \otimes I_3, \quad \sigma^B_{kl} = I_2 \otimes \sigma_{kl}.$$

The atomic operator $\sigma_{kl} = |k\rangle\langle l|$ takes an atom from the state $|l\rangle$ to the state $|k\rangle$. I_n is the $n \times n$ identity matrix. In the case of a two-level atom, the levels are described as $|0\rangle$ being the ground state and $|1\rangle$ as the excited state. For the three-level atom, the two excited levels are denoted by $|2\rangle$ and $|1\rangle$, while the ground state is denoted by $|0\rangle$. γ is the decay constant for the two level atom A, while γ_2 and γ_1 are the atomic decay constants of level $|2\rangle$ to level $|0\rangle$ and level $|1\rangle$ to level $|0\rangle$ for the three-level atom B, respectively (see Figure 6.2).

Let us consider a general density matrix with respect to the orthonormal basis $|1,2\rangle, |1,1\rangle, |1,0\rangle, \ldots, |0,0\rangle$,

$$\rho = \begin{pmatrix} \rho_{11} & \rho_{12} & \cdots & \rho_{16} \\ \rho_{21} & \rho_{22} & \cdots & \rho_{26} \\ \vdots & \vdots & \ddots & \vdots \\ \rho_{61} & \rho_{62} & \cdots & \rho_{66} \end{pmatrix}. \tag{6.20}$$

For this general density matrix there are 36 equations of motion derived from Eq. (6.18), 24 of them are uncoupled and can be solved easily. The remaining 12 equations are coupled.

6.1.4 Dynamical process of disentanglement

Dynamics of entanglement for pure states (Eq. (6.6))

For the density matrix of Eq. (6.6), the non-zero time evolved matrix elements are given by

$$\rho_{22}(t) = \beta^2 e^{-(\gamma+\gamma_1)t},$$

$$\rho_{26}(t) = \rho_{62}(t) = \pm \alpha \beta e^{\frac{-(\gamma+\gamma_1)t}{2}},$$

$$\rho_{33}(t) = \beta^2 \left(e^{-\gamma t} - e^{-(\gamma+\gamma_1)t} \right),$$

$$\rho_{55}(t) = \beta^2 \left(e^{-\gamma_1 t} - e^{-(\gamma+\gamma_1)t} \right),$$

$$\rho_{66}(t) = 1 - \beta^2 \left(e^{-\gamma t} + e^{-\gamma_1 t} - e^{-(\gamma+\gamma_1)t} \right).$$

The corresponding *negativity* is given by

$$N_1(\beta) = \max\left\{ 0, e^{-(\gamma+\gamma_1)t} \left[\beta^2 \left(2 - e^{\gamma t} - e^{\gamma_1 t} \right) + \sqrt{\beta^4 \left(e^{2\gamma t} + e^{2\gamma_1 t} \right) + (4\beta^2 - 6\beta^4) e^{(\gamma+\gamma_1)t}} \right] \right\}. \tag{6.21}$$

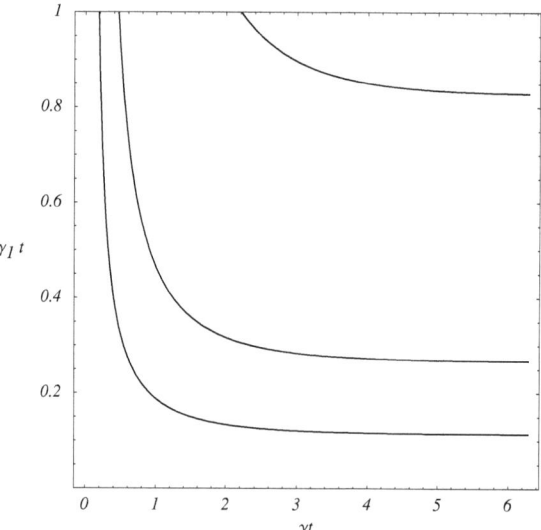

Figure 6.3: Time dependence of *negativity* Eq. (6.21), for different values of $\beta \in [\frac{1}{\sqrt{2}}, 1]$. The lower, middle and upper curves are for $\beta = 0.95, 0.9$ and 0.8 respectively. Each curve is a boundary between entangled and separable states. The states above and on the boundary are separable.

At $t = 0$, this *negativity* is given by $2\alpha\beta$. For maximally entangled states, i.e., $\alpha = \beta = 1/\sqrt{2}$, Eq. (6.21) reduces to

$$N_1 = e^{-(\gamma + \gamma_1)t}. \tag{6.22}$$

Hence maximally entangled state looses their entanglement at infinity. For the parameter range of $0 < \beta \leq 1/\sqrt{2}$, the states loose their entanglement in infinite time, however for $1/\sqrt{2} < \beta < 1$, the phenomenon of sudden death appears. This example is equivalent to pure states of two qubits [83]. Figure 6.3 shows contour plots of Eq. (6.21) for various values of β.

Disentanglement of pure states (Eq. (6.7))

Let us consider the density matrix of Eq. (6.7). This state is locally equivalent to state (6.6), and therefore has the same degree of entanglement.

After the interaction with both reservoirs, its non-zero matrix elements are given by

$$\rho_{11}(t) = \beta^2 e^{-(\gamma+\gamma_2)t},$$

$$\rho_{15}(t) = \rho_{51}(t) = \pm\alpha\beta e^{\frac{-(\gamma+\gamma_1+\gamma_2)t}{2}},$$

$$\rho_{33}(t) = \beta^2\left(e^{-\gamma t} - e^{-(\gamma+\gamma_2)t}\right),$$

$$\rho_{44}(t) = \beta^2\left(e^{-\gamma_2 t} - e^{-(\gamma+\gamma_2)t}\right),$$

$$\rho_{66}(t) = 1 - \alpha^2\left(e^{-\gamma_1 t} - \beta^2\left(e^{-\gamma t} + e^{-\gamma_2 t} - e^{-(\gamma+\gamma_2)t}\right)\right). \quad (6.23)$$

The appearance of decay factors γ_1 and γ_2, simply reflects the fact that both upper levels of the three-level atom are now involved. As we discussed earlier, the transitions $|2\rangle \to |0\rangle$ and $|1\rangle \to |0\rangle$ may cause *quantum interference*. This interference has a profound effect on the process of disentanglement in systems of two entangled qutrits [87]. In our qubit-qutrit systems, it also affects the process of disentanglement in a similar manner. For $k = 1$, there is no quantum interference. As k decreases, quantum interference increases and it is maximum for $k = 0$.

Negativity of the time evolved state is given by

$$N_2 = e^{-\gamma_2 t}\left\{\beta^2(e^{-\gamma t} - 1) + \sqrt{\beta^4(-1 + e^{-\gamma t})^2 + 4\alpha^2\beta^2 e^{-(\gamma+\gamma_1-\gamma_2)t}}\right\}. \quad (6.24)$$

We note that for $\alpha = \beta = 1/\sqrt{2}$ this *negativity* has the maximum value of 1 at $t = 0$, and states become separable at $t = \infty$. Nevertheless the process of disentanglement is different from Eq. (6.21). Sudden death never occurs for any value of β. However, quantum interference may be used to control the process of disentanglement. Figure 6.4 shows the behavior of *negativity* (Eq. (6.24)) for zero and maximum interference. The locally equivalent pure state $|\Phi_2'\rangle = \alpha|02\rangle + \beta|11\rangle$ exhibits the same dynamics.

We have shown that certain pure entangled states for a given degree of entanglement exhibit sudden death of entanglement, while other locally equivalent pure states do not. However, quantum interference is an additional feature of higher dimensions of the Hilbert spaces, which can control the dynamics of entanglement.

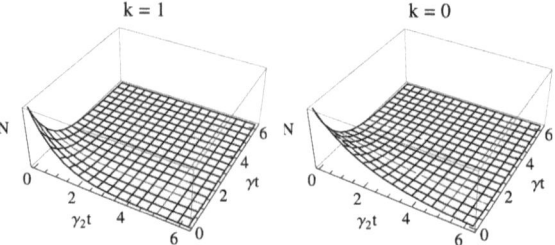

Figure 6.4: Time dependence of *negativity* for a maximally entangled state for zero and maximum interference, i.e., $k = 1$ and 0, respectively.

Disentanglement of mixed states

In this section, we consider an important class of mixed states of $2 \otimes N$ systems [142]. It has been shown that an arbitrary state ρ of $2 \otimes 3$ systems can be transformed to a state of the form in Eq. (6.25) by local unitary transformations. These states form a two-parameter class of states. For qubit-qutrit systems, they are given by

$$\rho_{a,c} = c|\Psi^-\rangle\langle\Psi^-| + b(|\Psi^+\rangle\langle\Psi^+| + |0,0\rangle\langle0,0| + |1,1\rangle\langle1,1|)$$
$$+ a(|0,2\rangle\langle0,2| + |1,2\rangle\langle1,2|), \qquad (6.25)$$

where $|\Psi^\pm\rangle = 1/\sqrt{2}(|0,1\rangle \pm |1,0\rangle)$, and the unit trace constrains the parameters to satisfy the relation

$$2a + 3b + c = 1.$$

If the state $\rho_{a,c}$ interacts with statistically independent reservoirs at zero-temperature, its time evolved non-zero matrix elements are given by

$$\rho_{11}(t) = a\,e^{-(\gamma+\gamma_2)t},$$

$$\rho_{22}(t) = b\,e^{-(\gamma+\gamma_1)t},$$

$$\rho_{33}(t) = \frac{e^{-\gamma t}}{2}\left(1 - 2\left(b\,e^{-\gamma_1 t} + a\,e^{-\gamma_2 t}\right)\right),$$

$$\rho_{35}(t) = \rho_{53}(t) = \left(\frac{b-c}{2}\right) e^{\frac{-(\gamma+\gamma_1)t}{2}},$$

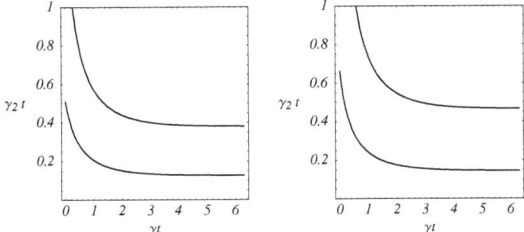

Figure 6.5: Contour plots of *negativity* is plotted for the fixed parameter $b = 0.02$. The lower and upper curves correspond to $c = 0.15$ and 0.2, respectively. The left graph is for $k = 1$ and the right one for $k = 0$. The region above and on each curve is that of sudden death of entanglement.

$$\rho_{44}(t) = a\left(2\,e^{-\gamma_2 t} - e^{-(\gamma+\gamma_2)t}\right),$$

$$\rho_{55}(t) = \left(\frac{3b+c}{2}\right) e^{-\gamma_1 t} - b\,e^{-(\gamma+\gamma_1)t},$$

$$\begin{aligned}\rho_{66}(t) &= 1 + \frac{e^{-(\gamma+\gamma_1+\gamma_2)t}}{2}\{e^{\gamma_1 t}(2a - 4a\,e^{\gamma t}) + e^{\gamma_2 t}(2b \\ &\quad +(-3b-c)\,e^{\gamma t} - e^{\gamma_1 t})\}.\end{aligned} \quad (6.26)$$

The expression of *negativity* in this case is lengthy. Rather than reproducing it here, we present some main results in diagrams.

Let us fix the parameter b in the above matrix elements and study sudden death in this class of mixed states. For $b = 0.02$, sudden death occurs in the range $0 < c \lesssim 0.302$ provided that interference is zero, that is, $\gamma_1 \approx \gamma_2$. However, sudden death is delayed when interference increases and for maximum interference, sudden death occurs in the range $0 < c \lesssim 0.2775$. Figure 6.5 shows contour plots of *negativity* versus dissipation factors γ_2 and γ for different values of the parameter c.

Similarly for the fixed parameter $b = 0.06$ and zero interference, sudden death occurs in the range $0 < c \lesssim 0.5493$. However, interference delays sudden death and for maximum interference sudden death occurs in the range $0 < c \lesssim 0.46295$. Figure 6.6 shows contour plots of *negativity* for different values of the parameter c.

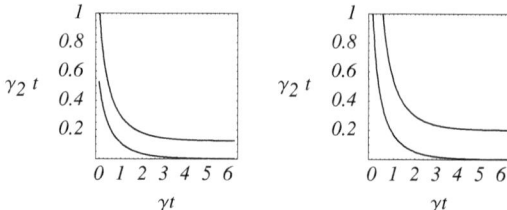

Figure 6.6: Same as Figure 6.5 for $b = 0.06$. The lower and upper curves are for $c = 0.25$ and 0.4, respectively.

6.2 Delaying, hastening and avoiding sudden death

In this section, we discuss the possibility of accelerating, delaying, or avoiding sudden death of entanglement. For this purpose, we define a single parameter class of mixed states. These states can be defined for all dimensions of the Hilbert spaces. This class of states is defined by the equation

$$\rho_a = \frac{1}{2}\{\,|\Omega\rangle\langle\Omega| + a\,|\omega_1\rangle\langle\omega_1| + (1-a)\,|\omega_2\rangle\langle\omega_2|\,\}, \qquad (6.27)$$

where $0 \leq a \leq 1$, $|\omega_1\rangle = |1,2\rangle$, and $|\omega_2\rangle = |0,0\rangle$ are separable states. The state $|\Omega\rangle$ is a maximally entangled pure state chosen so that it is orthogonal to both $|\omega_1\rangle$ and $|\omega_2\rangle$. Let us apply local unitary transformations, denoted by U_A and U_B (U_A and U_B denote the transformations from the computational basis to the Schmidt basis on the qubit and the qutrit respectively), the density matrix is transformed as

$$\tilde{\rho}_a = \frac{1}{2}\{\,|\tilde{\Omega}\rangle\langle\tilde{\Omega}| + a\,|\tilde{\omega}_1\rangle\langle\tilde{\omega}_1| + (1-a)\,|\tilde{\omega}_2\rangle\langle\tilde{\omega}_2|\,\}, \qquad (6.28)$$

with state $|\tilde{\Omega}\rangle = (U_A \otimes U_B)|\Omega\rangle$, $|\tilde{\omega}_1\rangle = (U_A \otimes U_B)|\omega_1\rangle$ and $|\tilde{\omega}_2\rangle = (U_A \otimes U_B)|\omega_2\rangle$. The state $|\tilde{\Omega}_1\rangle$ is again a maximally entangled pure state [92]. Local unitary transformations leave invariant the trace, the eigenvalues, and the degree of entanglement for a given density matrix. However, the subsequent evolution of entanglement after such actions may be very different [75, 102].

6.2.1 Effect of local unitary operations on entanglement dynamics

Our motivation is to take a state ρ and to analyze its dynamics before and after local actions. Some possible pure maximally entangled state vectors orthogonal to both $|w_1\rangle$ and $|w_2\rangle$ are given by

$$|\Omega_1\rangle = \frac{1}{\sqrt{2}}(|0,1\rangle \pm |1,0\rangle). \tag{6.29}$$

$$|\Omega_2\rangle = \frac{1}{\sqrt{2}}(|0,2\rangle \pm |1,0\rangle). \tag{6.30}$$

$$|\Omega_3\rangle = \frac{1}{\sqrt{2}}(|0,2\rangle \pm |1,1\rangle). \tag{6.31}$$

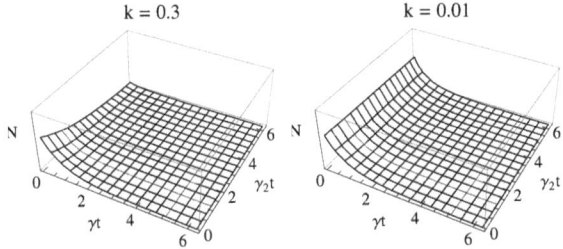

Figure 6.7: Evolution of *negativity* for Eq. (6.27) replacing state $|\Omega\rangle$ by state $|\Omega_1\rangle$ for two values of $k = \gamma_1/\gamma_2$ and for $a = 0.5$.

Case 1: Let us consider Eq. (6.27) with the state $|\Omega\rangle$ replaced by the state $|\Omega_1\rangle$. The solution of the master equation (6.18) and the eigenvalues of the partially transposed time evolved matrix can be found analytically. However, the resulting expressions for *negativity* are quite lengthy. Therefore we only present some results on the evolution of *negativity*. The amount of entanglement is maximum for $a = 1$ and minimum for $a = 0$. However sudden death never occurs for any value of the parameter a. Figure 6.7 describes the behavior of *negativity* for this class of states for two values of

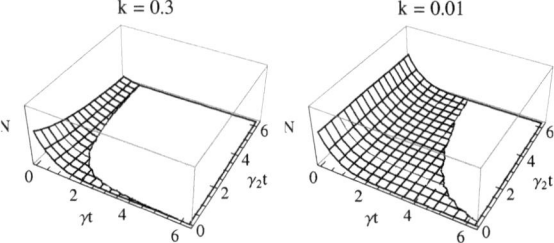

Figure 6.8: Effect of local switch "$0 \Leftrightarrow 1$" to state of Figure 6.7. The plots are for two values of k and for $a = 0.5$.

interference parameter k. Sudden death never occurs even for any value of a and k.

To see the effects of local actions, we consider the case of applying the switch "$0 \Leftrightarrow 1$" only at the qubit end. *Negativity* is plotted against the decay parameters after the switch in Figure 6.8, which shows that the larger the interference effects the slower is the disentanglement process. However, sudden death appears for some range of the parameter a, while for the locally equivalent state of Figure 6.7 it does not.

Case 2: Consider Eq. (6.27) with state $|\Omega\rangle$ replaced by state $|\Omega_2\rangle$. *Negativity* is plotted for these states in Figure 6.9. It is evident that for $a = 1$, the states loose their entanglement in a finite time, while for $a = 0$, they do so only at infinity. It is also interesting to observe that the states with $a = 1$ and $a = 0$ are related to each other by local switching (unitaries) at both ends, i.e., if Alice applies the switch "$0 \Leftrightarrow 1$" and Bob applies the switch "$0 \Leftrightarrow 2$", then $|\Omega_2\rangle$ remains invariant and $|1, 2\rangle$ is converted to $|0, 0\rangle$. Hence we observe again the situation, where it is possible to completely avert sudden death of entanglement by local unitary actions. This set of states has symmetry in its entanglement. The states are separable only at $a = 0.5$ and have an equal amount of entanglement for $a = 1$ and $a = 0$.

Case 3: We consider Eq. (6.27) with state $|\Omega\rangle$ replaced by state $|\Omega_3\rangle$. These states have minimum entanglement for $a = 1$ and maximum entanglement for $a = 0$. The two extreme states with $a = 1$ and with $a = 0$ are not related by local unitary transformations. Local actions can change their evolution of entanglement considerably. Sudden death happens for these states in the range of $0.502 \lesssim a \leq 1$ for given values of k but for $0 \leq a \lesssim 0.502$ it does not. Figure 6.10 shows *negativity* for these states by

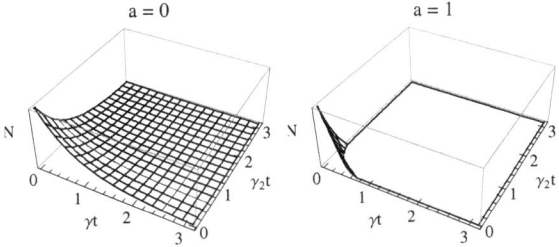

Figure 6.9: Evolution of *negativity* for an initially mixed state with $|\Omega\rangle \to |\Omega_2\rangle$ as in Eq. (6.30) for two values of a.

taking $a = 0.6$, $k = 0.3$, and 0.01.

Let us apply the local switch "0 ⇔ 1" to these states only at the qubit end. The resulting state changes its dynamics considerably and sudden death is avoided completely. Figure 6.11 shows *negativity* for $a = 0.6$ and for the same values of the parameter k after the application of a local switch.

6.2.2 Werner-like states

Let us consider now Werner-like states. This class has the property that under maximum interference condition, all states except a particular state, do not exhibit sudden death. Hence this family is quite robust against decoherence. Sudden death in these particular states can be avoided by local actions taken at only qubit end. This family of states is defined

$$\rho_\alpha = \alpha|\Phi\rangle\langle\Phi| + \frac{1-\alpha}{6}I. \qquad (6.32)$$

where $0 \leq \alpha \leq 1$ and $|\Phi\rangle$ is a maximally entangled pure state. I is the 6×6 identity matrix. This family is a single-parameter class of states and in the range $0 \leq \alpha \leq \frac{1}{4}$ these states are separable (PPT) and otherwise entangled. These states have the property that they maintain their basic structure under all local unitaries. In our particular environmental model these states also keep their density matrix structure in the master equation (6.18). As we could apply local unitary transformations denoted by U_A and U_B, the density matrix is transformed as $\tilde{\rho} = \alpha|\tilde{\Phi}\rangle\langle\tilde{\Phi}| + \frac{1-\alpha}{6}I$ with $|\tilde{\Phi}\rangle = (U_A \otimes U_B)|\Phi\rangle$. Again $|\tilde{\Phi}\rangle$ is a maximally entangled pure state if and only if $|\Phi\rangle$ is maximally entangled [92].

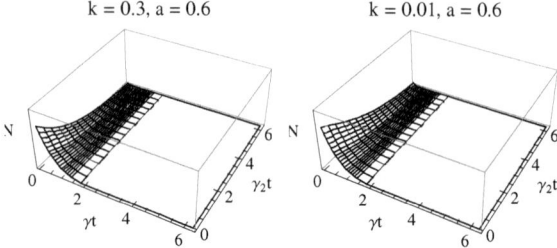

Figure 6.10: *Negativity* is plotted for Eq. (6.27) with $|\Omega\rangle$ replaced with Eq. (6.31) for two values of k and for $a = 0.6$.

Case 1: We first consider a pure maximally entangled state with the qutrit having both of its excited levels entangled with the qubit. Its density matrix is given by

$$\rho_1 = \alpha |\Phi_1\rangle\langle\Phi_1| + \frac{1-\alpha}{6} I, \qquad (6.33)$$

where $|\Phi_1\rangle = \frac{1}{\sqrt{2}}(|0,1\rangle \pm |1,2\rangle)$. Figure 6.12 depicts *negativity* for these states in the presence of quantum interference ($k < 1$). Figure 6.12 shows that for this family of states sudden death always happens in the range $0.25 < \alpha \lesssim 0.4$ but in the range $0.4 < \alpha \leq 1$ these states loose their entanglement asymptotically.

Consider a quantum state obtained by applying local unitary transformation "$0 \Leftrightarrow 1$" to Eq. (6.33) only at the qubit end. This transformation obviously leaves the identity matrix invariant. After the transformation the resulting states are given by

$$\rho_2 = \alpha |\Phi_2\rangle\langle\Phi_2| + \frac{1-\alpha}{6} I, \qquad (6.34)$$

where $|\Phi_2\rangle = \frac{1}{\sqrt{2}}(|1,1\rangle \pm |0,2\rangle)$. Figure 6.13 shows the corresponding *negativity*. Figures 6.12 and 6.13 look similar because the local switch has no considerable effect on the subsequent evolution of entanglement. For $\alpha > 0.4$, these states loose their entanglement only at infinity.

Case 2: We now slightly change the initial conditions and consider the case in which only one excited level of the three level atom is involved in pure entangled state, e.g. $|\Phi_3\rangle = \frac{1}{\sqrt{2}}(|0,0\rangle \pm |1,1\rangle)$. The corresponding

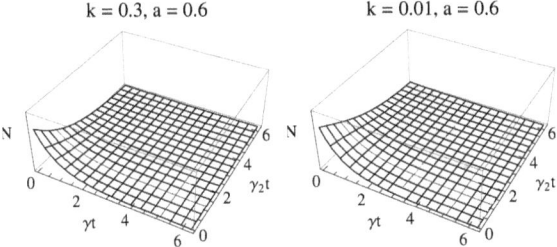

Figure 6.11: *Negativity* is plotted under the effect of local switch for the cases of Figure 6.10

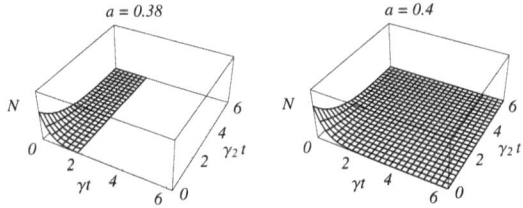

Figure 6.12: Evolution of *negativity* for Eq. (6.32) replacing state $|\Omega\rangle$ by state $|\Phi_1\rangle$ for two values of α and for $k = 0.3$.

mixed states are given by

$$\rho_3 = \alpha|\Phi_3\rangle\langle\Phi_3| + \frac{1-\alpha}{6}I. \qquad (6.35)$$

Figure 6.14 shows the dynamics of *negativity* for the initial state of Eq. (6.35). From Figure 6.14, it seems that these states are less robust compared with the states of Eq. (6.33) as sudden death phenomenon is extended up to the range $0.25 < \alpha < 1$ with $k = 0.3$. However, for the particular case of maximum interference, sudden death occurs only in the range $0.25 < \alpha \lesssim 0.5189$. In the range $0.519 \lesssim \alpha \leq 1$ these states always decay asymptotically.

Now we consider states related to Eq. (6.35) by a local unitary switch "$0 \leftrightarrow 1$" applied only at the qubit end. The density matrix after the switch

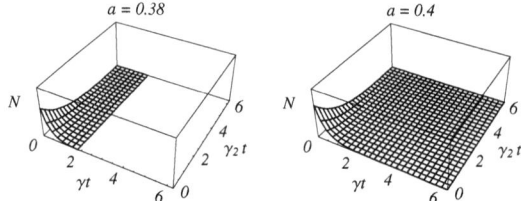

Figure 6.13: Evolution of *negativity* for Eq. (6.34) for two values of α and for $k = 0.3$.

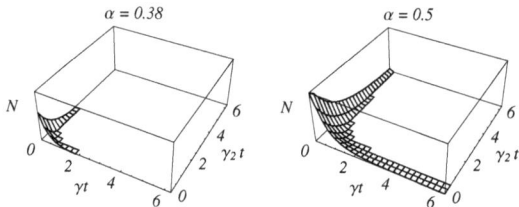

Figure 6.14: *Negativity* is plotted for Eq. (6.35) for two values of α and for $k = 0.3$.

is given by

$$\rho_4 = \alpha |\Phi_4\rangle\langle\Phi_4| + \frac{1-\alpha}{6} I , \tag{6.36}$$

where $|\Phi_4\rangle = \frac{1}{\sqrt{2}}(|1,0\rangle \pm |0,1\rangle)$. Figure 6.15 shows the corresponding *negativity* for these states. The effect of the local switch is quite clear in this case. For $k = 0.3$ sudden death always happens in the range $0.25 < \alpha \lesssim 0.55$ but in the range $0.55 \lesssim \alpha \leq 1$, the states decay asymptotically. For the case of maximum interference, sudden death always happens in the range $0.25 < \alpha \lesssim 0.4606$ but in the range $0.4606 \lesssim \alpha \leq 1$, the states decay asymptotically.

Case 3: Finally we consider the case where only level 2 for the qutrit is involved in pure entangled state. The states are given by

$$\rho_5 = \alpha |\Phi_5\rangle\langle\Phi_5| + \frac{1-\alpha}{6} I , \tag{6.37}$$

where $|\Phi_5\rangle = \frac{1}{\sqrt{2}}(|0,0\rangle \pm |1,2\rangle)$. Figure 6.16 shows *negativity* for Eq. (6.37). These states are the most fragile states because finite-time disentanglement

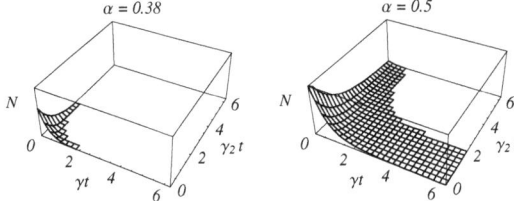

Figure 6.15: Evolution of *negativity* for an initially mixed state Eq. (6.36) for two values of α and for $k = 0.3$.

always happens for these states in the range $0.25 < \alpha < 1$ even in the presence of maximum interference. It is surprising that if the doubly excited component $|1,2\rangle$ is entangled with $|0,0\rangle$ in pure state, no matter how small the amount of maximally mixed state is added into it, sudden death appears. On the other hand if doubly excited component is entangled with $|0,1\rangle$ (Eq. (6.33)), sudden death does not appear (for $\alpha > 0.4$). Eq. (6.37) is invariant under the local switching, "$0 \Leftrightarrow 1$" and "$0 \Leftrightarrow 2$" at qubit and qutrit ends respectively. Hence for this set, local switching only at one end is effective.

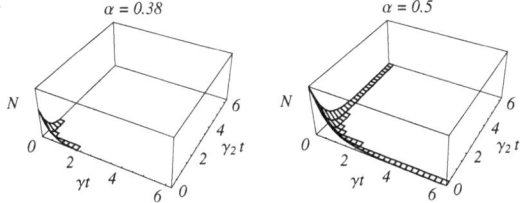

Figure 6.16: Dynamics of *negativity* for an initially mixed state Eq. (6.37) for two values of α and for $k = 0.3$.

Let us apply the local switch "$0 \Leftrightarrow 1$" only at the qubit end. So, the locally equivalent states to Eq. (6.37) are given by

$$\rho_6 = \alpha |\Phi_6\rangle\langle\Phi_6| + \frac{1-\alpha}{6} I, \qquad (6.38)$$

where $|\Phi_6\rangle = \frac{1}{\sqrt{2}}(|1,0\rangle \pm |0,2\rangle)$. Figure 6.17 shows the corresponding *negativity*. Sudden death of entanglement for these states always occurs in

85

the range $0.25 < \alpha \lesssim 0.544$ for $k = 0.3$. For the special case of maximum interference sudden death happens for $0.25 < \alpha \lesssim 0.5189$. For $0.519 \lesssim \alpha \leq 1$ these states decay asymptotically. Therefore, we can conclude that these states loose their entanglement only at infinity. Hence simple local actions can avoid sudden death of entanglement in these states.

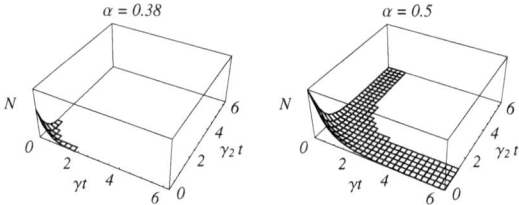

Figure 6.17: *Negativity* is plotted for an initially mixed state Eq. (6.38) for two values of α and for $k = 0.3$.

6.2.3 Avoiding finite-time disentanglement during the interaction process

In the previous section, we have discussed some cases, in which local unitary actions were taken at $t = 0$, i.e., before the interaction starts between a principal system and an environment. In this section, we explore the freedom in time to apply transformations in such a way that sudden death is delayed or avoided completely. We will start with an initial condition, which necessarily leads to sudden death. We apply local actions during time evolution and demonstrate the phenomenon of delaying or avoiding sudden death. Depending upon the time t_{sw} such actions are taken, sudden death can be delayed or even averted altogether. For this purpose we fix the parameter γ_2 equal to γ and relate the time of switching t_{sw} to the time of sudden death t_{end}. In general $\gamma \neq \gamma_2$ but to conveniently picturize the delay or avoidance of sudden death, we take them equal for this case.

Let us consider the case in which the doubly excited component $|\omega_1\rangle$ is mixed with $|\Omega_2\rangle$ (Eq. (6.30)) i.e., Eq. (6.27) with $a = 1$. It was shown in the previous section (Figure 6.9) that such states exhibit sudden death. We can apply local switches "$0 \Leftrightarrow 1$" and "$0 \Leftrightarrow 2$" at both ends. This switch can only exchange the two extreme elements on the main diagonal of our density matrix i.e., "$\rho_{11} \Leftrightarrow \rho_{66}$". Depending upon the time of switching t_{sw}, the time of sudden death t_{end} can be increased up to infinity. Figure 6.18

shows the relation between t_{sw} and t_{end}. Hence any switch made before $t_{sw} \approx 0.2318/\gamma$ delays sudden death. A switch made before $t_{sw} \approx 0.0985/\gamma$ completely avoids sudden death and we have only asymptotic decay.

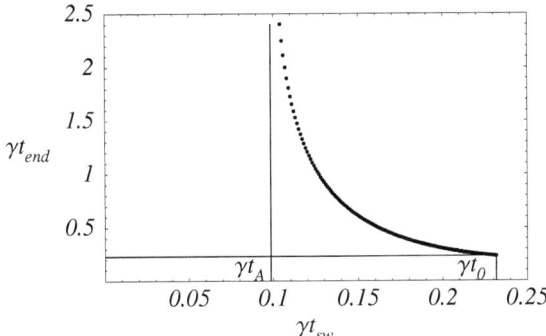

Figure 6.18: The time for the end of entanglement t_{end} is plotted against the time of switching t_{sw}. At switch time of $t_0 \approx 0.2318/\gamma$, the curve rises slowly to infinite time at $t_A \approx 0.0985/\gamma$.

This example is similar to qubit-qubit systems, as there is only one excited level involved in the three level atom. However there is one remarkable difference between the case studied in Section 5.1 (Ref. [102]) and Figure 6.18. In Figure 5.2 of Section 5.1, there is a dip (hastening) of *negativity* before rising rapidly to infinity. In contrast, there is no dip in Figure 6.18 and sudden death is always delayed with switching and finally rising asymptotically. In fact, although in the current case the quantum state is similar to the case studied in Section 5.1, the initial amount of entanglement in both cases is different. Hence, surprisingly the amount of entanglement present in a quantum state is also responsible for the future trajectory of its measure.

Now we consider the case 3 of the previous section as our initial state and apply local switch "0 ⇔ 1" only at the qubit end. In this specific example both excited levels of the three level atom are involved along with interference feature. We set the parameter $k = 0.1$ and in order to show the relation between t_{end} and t_{sw}, we take γ_2 equal to γ. The result of such a switch is to exchange $\rho_{11} \Leftrightarrow \rho_{44}$, $\rho_{22} \Leftrightarrow \rho_{55}$, $\rho_{33} \Leftrightarrow \rho_{66}$, and $\rho_{24} \Leftrightarrow \rho_{51}$. Figure 6.19 shows the relation between t_{end} and t_{sw}. Hence such a switch applied before $\gamma t_{sw} \approx 0.279$ leads to asymptotic decay of entanglement.

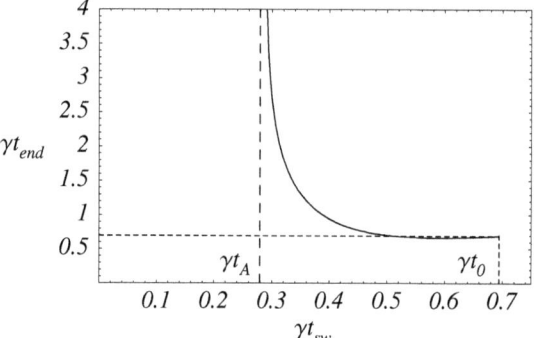

Figure 6.19: The time for the end of entanglement t_{end} is plotted against the time of switching t_{sw}. At switch time of $t_0 \approx 0.694/\gamma$ the curve rises slowly to infinite time at $t_A \approx 0.279/\gamma$.

It is sufficient to demonstrate this possibility of avoiding sudden death by local actions implemented during the system-environment interaction process in these two cases. Similar analysis can be carried out for all other cases mentioned in this Chapter.

6.3 Asymptotic states and interference

The dynamics of a given quantum state ρ by Eq. (6.18) depends significantly on quantum interference as characterized by the parameter $k = \gamma_1/\gamma_2$. For $\gamma_1 > 0$, whatever the initial state, whether pure or mixed, and whether entangled or separable, the final state reached in asymptotic time is clearly both of atoms in their ground states $|0,0\rangle$.

For maximum interference $\gamma_1 = 0$ the final state depends on the initial state. If ρ_{ij} denotes a matrix element of the general quantum state, the asymptotic solution of Eq. (6.18) leads to the following four possible non-zero asymptotic matrix elements

$$\begin{aligned}
\rho_{55}(\infty) &= \rho_{22} + \rho_{55}, \\
\rho_{56}(\infty) &= \rho_{23} + \rho_{56}, \\
\rho_{65}(\infty) &= \rho_{32} + \rho_{65}, \\
\rho_{66}(\infty) &= \rho_{11} + \rho_{33} + \rho_{44} + \rho_{66}.
\end{aligned} \quad (6.39)$$

Hence whatever the values of these four matrix elements, the asymptotic states $\rho(t = \infty)$ are definitely separable (non-entangled). This is in contrast with qutrit-qutrit systems, where some of the asymptotic states can be entangled [87]. Hence, quantum interference present only in qutrit part of qubit-qutrit entangled system is not enough for the possibility of having asymptotic entangled states. The reduced density matrix for the qubit only is

$$\rho_A(\infty) = \begin{pmatrix} 0 & 0 \\ 0 & 1 \end{pmatrix}, \qquad (6.40)$$

whereas the reduced density matrix for the qutrit is given by

$$\rho_B(\infty) = \begin{pmatrix} 0 & 0 & 0 \\ 0 & \rho_{55}(\infty) & \rho_{56}(\infty) \\ 0 & \rho_{65}(\infty) & \rho_{66}(\infty) \end{pmatrix}. \qquad (6.41)$$

The qutrit part always remains in a superposition of ground and first excited level, i.e., $(|0\rangle + |1\rangle)$. In a qutrit-qutrit system, both reduced density matrices are always of the form of Eq. (6.41). Hence we expect that in bipartite $M \otimes N$-systems, the possibility of asymptotically entangled states in the presence of maximum interference is only possible if at least $M, N \geq 3$. Thus qutrit-qutrit systems are the minimal dimensional systems with such a possibility. Both qubit-qubit and qubit-qutrit states always end up with separable asymptotic states in this particular model.

6.4 Sudden death of qubit-qutrit systems by phase damping

In this section we summarize some of the results of the recent investigation on phase damping [100]. Let the general density matrix of the composite qubit-qutrit system be given by $\rho_{AR} = [\rho_{ij}]$, where $\rho_{ij}^* = \rho_{ji}$, $\sum_i \rho_{ii} = 1$, with $i, j = 1, \ldots, 6$. This matrix contains the terms responsible for subsystem coherence and joint-system coherence. The joint-system coherent terms are associated with quantum entanglement. The individual subsystem can be represented by a reduced density matrix by tracing out the variables of other subsystem. The general reduced density matrix ρ_A for the qubit is given by

$$\rho_A = \begin{pmatrix} (\rho_{11} + \rho_{22} + \rho_{33}) & (\rho_{14} + \rho_{25} + \rho_{36}) \\ (\rho_{41} + \rho_{52} + \rho_{63}) & (\rho_{44} + \rho_{55} + \rho_{66}) \end{pmatrix}. \qquad (6.42)$$

Quantum states in which the subsystems are incoherent can be those in which the composite system possesses high joint-state coherence and are entangled. Therefore to study the effect of local dephasing noise on entanglement, we take all off-diagonal terms in the reduced density matrices to be zero. Therefore the incoherent reduced qubit state is given by

$$\tilde{\rho}_A = \begin{pmatrix} (\rho_{11} + \rho_{22} + \rho_{33}) & 0 \\ 0 & (\rho_{44} + \rho_{55} + \rho_{66}) \end{pmatrix}. \quad (6.43)$$

Similarly, the reduced density matrix for the qutrit ρ_B is given by

$$\rho_B = \begin{pmatrix} (\rho_{11} + \rho_{44}) & (\rho_{12} + \rho_{45}) & (\rho_{13} + \rho_{46}) \\ (\rho_{21} + \rho_{54}) & (\rho_{22} + \rho_{55}) & (\rho_{23} + \rho_{56}) \\ (\rho_{31} + \rho_{64}) & (\rho_{32} + \rho_{65}) & (\rho_{33} + \rho_{66}) \end{pmatrix}, \quad (6.44)$$

with the incoherent qutrit reduced state

$$\tilde{\rho}_B = \begin{pmatrix} (\rho_{11} + \rho_{44}) & 0 & 0 \\ 0 & (\rho_{22} + \rho_{55}) & 0 \\ 0 & 0 & (\rho_{33} + \rho_{66}) \end{pmatrix}. \quad (6.45)$$

Composite density matrices that yield reduced states of the form $\tilde{\rho}_A$ and $\tilde{\rho}_B$ are given by

$$\tilde{\rho}_{AB} = \begin{pmatrix} \rho_{11} & 0 & 0 & 0 & \rho_{15} & \rho_{16} \\ 0 & \rho_{22} & 0 & \rho_{24} & 0 & \rho_{26} \\ 0 & 0 & \rho_{33} & \rho_{34} & \rho_{35} & 0 \\ 0 & \rho_{42} & \rho_{43} & \rho_{44} & 0 & 0 \\ \rho_{51} & 0 & \rho_{53} & 0 & \rho_{55} & 0 \\ \rho_{61} & \rho_{62} & 0 & 0 & 0 & \rho_{66} \end{pmatrix}. \quad (6.46)$$

For simplicity, consider the specific one-parameter class of states given by

$$\tilde{\rho}_{AB}(x) = \begin{pmatrix} \frac{1}{4} & 0 & 0 & 0 & 0 & x \\ 0 & \frac{1}{8} & 0 & 0 & 0 & 0 \\ 0 & 0 & \frac{1}{8} & 0 & 0 & 0 \\ 0 & 0 & 0 & \frac{1}{8} & 0 & 0 \\ 0 & 0 & 0 & 0 & \frac{1}{8} & 0 \\ x & 0 & 0 & 0 & 0 & \frac{1}{4} \end{pmatrix}, \quad (6.47)$$

where $0 \leq x \leq 1/4$.

Noise model

Let us consider local dephasing noise acting on subsystems that are no longer interacting with each other. The most general time evolved density matrix in the operator-sum representation can be written as

$$\rho(t) = \varepsilon(\rho(0)) = \sum_\mu K_\mu(t)\rho(0)K_\mu^\dagger(t), \qquad (6.48)$$

where the linear operators $K_\mu(t)$ satisfy a completeness relation given by $\sum_\mu K_\mu^\dagger K_\mu = I$ and guarantee that the evolution is trace-preserving. It represents the influence of noise which may be global or local. For local and multi-local dephasing, the operators $K_\mu(t)$ are of the form $K_\mu(t) = F_j(t) \otimes E_i(t)$, so that

$$\rho_{AB}(t) = \varepsilon(\rho(0)) = \sum_{i=1}^{2}\sum_{j=1}^{3} F_j(t) \otimes E_i(t)\rho_{AB}(0)E_i^\dagger(t) \otimes F_j^\dagger(t), \qquad (6.49)$$

where

$$\begin{aligned}
E_1(t) &= \mathrm{diag}(1,\gamma_A) \otimes \mathrm{diag}(1,1,1) = \mathrm{diag}(1,1,1,\gamma_A,\gamma_A,\gamma_A), &(6.50)\\
E_2(t) &= \mathrm{diag}(0,\omega_A) \otimes \mathrm{diag}(1,1,1) = \mathrm{diag}(0,0,0,\omega_A,\omega_A,\omega_A), &(6.51)\\
F_1(t) &= \mathrm{diag}(1,1) \otimes \mathrm{diag}(1,\gamma_B,\gamma_B) = \mathrm{diag}(1,\gamma_B,\gamma_B,1,\gamma_B,\gamma_B), &(6.52)\\
F_2(t) &= \mathrm{diag}(1,1) \otimes \mathrm{diag}(0,\omega_B,0) = \mathrm{diag}(0,\omega_B,0,0,\omega_B,0), &(6.53)\\
F_3(t) &= \mathrm{diag}(1,1) \otimes \mathrm{diag}(0,0,\omega_B) = \mathrm{diag}(0,0,\omega_B,0,0,\omega_B), &(6.54)
\end{aligned}$$

with

$$\begin{aligned}
\gamma_A(t) &= \exp\{-\Gamma_A t/2\}, \quad \gamma_B(t) = \exp\{-\Gamma_B t/2\},\\
\omega_A(t) &= \sqrt{1-\gamma_A^2(t)}, \quad \omega_B(t) = \sqrt{1-\gamma_B^2(t)},
\end{aligned} \qquad (6.55)$$

and $\mathrm{diag}(1,\gamma_A)$ is a diagonal matrix with elements 1 and γ_A, etc. The operators $E_i(t)$ and $F_j(t)$ individually satisfy the completeness relation, i.e. $\sum_i E_i^\dagger(t)E_i(t) = I$ and $\sum_j F_j^\dagger(t)F_j(t) = I$ and induce local dephasing in the qubit and the qutrit states, respectively. Now we can consider three noise situations. i.e., the qubit dephasing only, the qutrit dephasing only, and the combined local dephasing. In the first case, $F_j(t) = I$, and the time evolution of the density matrix (6.47) is given as

$$\tilde{\rho}_{AB}(x,t) = \begin{pmatrix} \frac{1}{4} & 0 & 0 & 0 & 0 & x\gamma_A \\ 0 & \frac{1}{8} & 0 & 0 & 0 & 0 \\ 0 & 0 & \frac{1}{8} & 0 & 0 & 0 \\ 0 & 0 & 0 & \frac{1}{8} & 0 & 0 \\ 0 & 0 & 0 & 0 & \frac{1}{8} & 0 \\ x\gamma_A & 0 & 0 & 0 & 0 & \frac{1}{4} \end{pmatrix}. \qquad (6.56)$$

Negativity for this matrix is given as

$$N(\tilde{\rho}_{AB}(x,t)) = \max\{0, x\gamma_A - 1/8\}. \tag{6.57}$$

This function is zero for a finite time. Hence the global coherence vanishes in a finite time, i.e., sudden death, while decoherence occurs only asymptotically in the large-time limit.

Similarly, for noise acting on the qutrit B alone, the time evolved density matrix is given as

$$\tilde{\rho}_{AB}(x,t) = \begin{pmatrix} \frac{1}{4} & 0 & 0 & 0 & 0 & x\gamma_B \\ 0 & \frac{1}{8} & 0 & 0 & 0 & 0 \\ 0 & 0 & \frac{1}{8} & 0 & 0 & 0 \\ 0 & 0 & 0 & \frac{1}{8} & 0 & 0 \\ 0 & 0 & 0 & 0 & \frac{1}{8} & 0 \\ x\gamma_B & 0 & 0 & 0 & 0 & \frac{1}{4} \end{pmatrix}. \tag{6.58}$$

In this case *negativity* is given as

$$N(\tilde{\rho}_{AB}(x,t)) = \max\{0, x\gamma_B - 1/8\}. \tag{6.59}$$

Hence sudden death of entanglement also takes place in this case of local dephasing noise acting on the qutrit alone. However, the full decoherence of the composite system occurs only in the large time limit, i.e., $t \to \infty$.

For the third situation where both subsystems are affected by local noise, the time evolved density matrix is given as

$$\tilde{\rho}_{AB}(x,t) = \begin{pmatrix} \frac{1}{4} & 0 & 0 & 0 & 0 & x\gamma_A\gamma_B \\ 0 & \frac{1}{8} & 0 & 0 & 0 & 0 \\ 0 & 0 & \frac{1}{8} & 0 & 0 & 0 \\ 0 & 0 & 0 & \frac{1}{8} & 0 & 0 \\ 0 & 0 & 0 & 0 & \frac{1}{8} & 0 \\ x\gamma_A\gamma_B & 0 & 0 & 0 & 0 & \frac{1}{4} \end{pmatrix}. \tag{6.60}$$

In this case *negativity* is given as

$$N(\tilde{\rho}_{AB}(x,t)) = \max\{0, x\gamma_A\gamma_B - 1/8\}. \tag{6.61}$$

By the same arguments, multi-local dephasing induces sudden death even more quickly than in the case of single local dephasing noise.

Summary of Chapter 6

In summary we have investigated the effects of local unitary actions on sudden death of entanglement in two classes of quantum states of qubit-qutrit systems. It has been demonstrated that such operations can completely avoid sudden death. We have also discussed the possibility of delaying or avoiding sudden death with local actions taken later when the dissipative dynamics has already degraded entanglement. We have shown that up to a critical time, sudden death can be delayed and if local actions are taken before that critical value of interaction time, sudden death can be avoided. The pursuit of robust quantum states, which can tolerate effects of noise and which do not undergo the peculiar phenomenon of sudden death, is vital for quantum information processing. In this regard we considered a single parameter class of states, namely Werner-like states, which are quite robust against sudden death. Except for a pair of quantum states, all states in this class do not undergo sudden death. Moreover, it is also possible to avoid sudden death in this pair by applying local action at the qubit end. We have also studied asymptotic states in the presence of maximum interference. We point out that while it is possible for qutrit-qutrit systems to have entangled asymptotic states, however, for qubit-qutrit system it is impossible. All entangled quantum states in qubit-qutrit systems, while interacting with vacuum at two distant locations, end up in separable states. We expect similar results for qubit-qudit systems.

Chapter 7
Summary and conclusion

This thesis is concerned with dynamics of quantum entanglement. Its main motivation is to stabilize entanglement against decoherence effects. Entanglement is a dynamical quantity and it is very important to know its time evolution in various physical systems. Since quantum systems interact with environments and this unwanted interaction leads to decoherence. Docoherence leads to decay of entanglement. Recently, a special type of decoherence is observed which attacks quantum entanglement only and finishes it in finite time. This type of decoherence is named as sudden death of entanglement. Sudden death of entanglement has been predicted for many physical systems. It has been observed in laboratory as well. We need to take some measures to prevent quantum systems from this effect. This thesis is an effort in this regard. In particular, this thesis focuses on controlling finite-time disentanglement of bipartite entangled states of qubit-qubit and qubit-qutrit systems.

In the theory of open quantum systems, dynamics of a principal system is obtained by taking trace over an environment. Under realistic conditions open quantum systems interact with environments and dynamics of the combined system is unitary. However, dynamics of open systems is not describable by unitary transformations due to their interactions with environments. The dynamics of open systems can be described by the *master equations* under certain circumstances. We have studied entanglement sudden death of two non-interacting qubits interacting with their own statistically independent reservoirs at zero- and finite-temperatures. We have shown that in zero-temperature reservoirs, sudden death may appear depending on the initial preparation of quantum states. We have analyzed a particular set of entangled states, namely X-states which preserve their form under time evolution. We have shown that all those X-states which exhibit sudden death can be converted to X-states which do not exhibit sudden death. We have shown that such conversion is possible not only before the interaction process but also during the interaction with the environment. However, there is always some critical time, after which it is no longer possible to avert finite-time disentanglement. Interestingly, we have observed that it is possible to hasten, delay and even avert sudden death of entanglement by local unitary transformations. However, the complete avertion of sudden death is only possible in zero-temperature reservoirs. In finite-temperature reservoirs, it is still possible to accelerate or delay finite-time disentanglement up to some finite time. We have analytically shown that all X-states exhibit sudden death in finite-temperature reservoirs. We have analyzed the set of unitary operations and found the conditions on local unitary transformations which map X-states to X-states.

We have enlarged the dimension of the Hilbert space to 6 and considered qubit-qutrit systems. We have studied dynamics of entanglement for various initial qubit-qutrit states. We have first provided the evidence of sudden death due to amplitude damping and phase damping. Then we have studied the possibility of delaying, hastening, and avoiding sudden death. Again it is possible to hasten, delay, and avoid finite-time disentanglement via local unitary transformations. We have studied qubit-qutrit systems interacting with two independent reservoirs at zero-temperature only. However, we remark that *quantum interference* is an additional feature which can control the disentanglement process. In contrast to qutrit-qutrit systems, all asymptotic states in qubit-qutrit systems are separable.

Bibliography

[1] E. Schrödinger, die Naturwissenschaften **23**, 807 (1935).

[2] A. Einstein, B. Podolsky, and N. Rosen, Phys. Rev. **47**, 777 (1935).

[3] N. Bohr, Phys. Rev. **48**, 696 (1935).

[4] J. S. Bell, Physics **1**(3), 195 (1964).

[5] J. F. Clauser, M. A. Horne, A. Shimony, and R. A. Holt, Phys. Rev. Lett. **23**, 880 (1969).

[6] J. F. Clauser and M. A. Horne, Phys. Rev. D **10**, 526 (1974).

[7] A. Aspect, P. Grangier, and G. Roger, Phys. Rev. Lett. **47**, 460 (1981).

[8] A. Aspect, P. Grangier, and G. Roger, Phys. Rev. Lett. **49**, 91 (1982).

[9] A. Aspect, J. Dalibard, and G. Roger, Phys. Rev. Lett. **49**, 1804 (1982).

[10] R. García-Patrón, J. Fiurášek, N. J. Cerf, J. Wenger, R. Tualle-Brouri, and P. Grangier, Phys. Rev. Lett. **93**, 130409 (2004).

[11] T. Scheidl, R. Ursin, J. Kofler, S. Ramelow, X-S Ma, T. Herbst, L. Ratschbacher, A. Fedrizzi, N. Langford, T. Jennewein, and A. Zeilinger, arXiv:quant-ph/0811.3129.

[12] C. H. Bennett and G. Brassard, in *Proceedings of the IEEE International conference on Computers, Systems and Signal Processing* (IEEE Computer Society Press, Bangalore, India, December 1984) pp. 175-179

[13] D. Deutsch, Proc. R. Soc. London, Ser. A **400**, 97 (1985).

[14] R. P. Feynman, Int. J. Theor. Phys. **21**, 467 (1982).

[15] P. W. Shor, Phys. Rev. A **52**, 2493 (1995).

[16] A. Steane, Phys. Rev. Lett. **77**, 793 (1996).

[17] A. K. Ekert, Phys. Rev. Lett. **67**, 661 (1991).

[18] C. H. Bennett and S. J. Wiesner, Phys. Rev. Lett. **69**, 2881 (1992).

[19] C. H. Bennett, G. Brassard, C. Crepeau, R. Josza, A. Peres, and W. K. Wootters, Phys. Rev. Lett. **70**, 1895 (1993).

[20] K. Mattle, H. Weinfurter, P. G. Kwiat, and A. Zeilinger, Phys. Rev. Lett. **76**, 4656 (1996).

[21] D. Bouwmeester, J. W. Pan, K. Mattle, M. Eibl, H. Weinfurter, and A. Zeilinger, Nature **390**, 575 (1997).

[22] D. Boschi, S. Branca, F. D. Martini, L. Hardy, and S. Popescu, Phys. Rev. Lett. **80**, 1121 (1998).

[23] J. W. Pan, D. Bouwmeester, H. Weinfurter, and A. Zeilinger, Phys. Rev. Lett. **80**, 3891 (1998).

[24] A. Furusawa, J. L. Sorensen, S. L. Braunstein, C. A. Fuchs, H. J. Kimble, and E. S. Polzik, Science **282**, 706 (1998).

[25] T. Jennewein, C. Simon, G. Weihs, H. Weinfurter, and A. Zeilinger, Phys. Rev. Lett. **54**, 4729 (2000).

[26] D. S. Naik, C. G. Peterson, A. G. White, A. J. Berglund, and P. G. Kwiat, Phys. Rev. Lett. **84**, 4733 (2000).

[27] W. Tittel, J. Brendel, H. Zbinden, and N. Gisin, Phys. Rev. Lett. **84**, 4737 (2000).

[28] H-K. Lo, T. Spiller, and S. Popescu, *Introduction to quantum computation and information* (World Scientific, 1999).

[29] M. A. Nielsen and I. L. Chuang, *Quantum Computation and Quantum Information* (Cambridge Univ. Press, Cambridge, 2000).

[30] D. Bouwmeester, A. K. Ekert, and A. Zeilinger, *The physics of quantum information: quantum cryptography, quantum teleportation, quantum computation* (Springer, New York, 2000).

[31] G. Alber, T. Beth, M. Horodecki, P. Horodecki, R. Horodecki, M. Rötteler, H. Weinfurter, R. Werner, and A. Zeilinger, *Quantum Information* (Springer-Verlag, Berlin, 2001).

[32] S. L. Braunstein, and A. K. Pati, *Quantum Information with Continuous Variables* (Kluwer Academics Publishers, 2003).

[33] D. Bruß and G. Leuchs, *Lectures on Quantum Information* (Wiley-Vch GmbH & Co. KGaA, 2007).

[34] R. Horodecki, P. Horodecki, M. Horodecki, and K. Horodecki, Rev. Mod. Phys. **81**, 865 (2009).

[35] L. Clarisse, PhD thesis, University of York, England, arXiv:quant-ph/0612072.

[36] R. F. Werner, Phys. Rev. A **40**, 4277 (1989).

[37] A. Peres, Phys. Rev. Lett. **77**, 1413 (1996).

[38] M. Horodecki, P. Horodecki, and R. Horodecki, Phys. Lett. A **223**, 1 (1996).

[39] P. Horodecki, M. Lewenstein, G. Vidal, and I. Cirac, Phys. Rev. A **62**, 032310 (2000).

[40] B. M. Terhal, Theo. Comp. Sc. **287**, 313 (2001).

[41] P. Horodecki, Theo. Comp. Sc. **293**, 589 (2003).

[42] M. Horodecki, P. Horodecki, and R. Horodecki, Phys. Rev. Lett. **80**, 5239 (1998).

[43] D. P. DiVincenzo, P. W. Shor, J. A. Smolin, B. M. Terhal, and A. V. Thapliyal, Phys. Rev. A **61**, 062312 (2000).

[44] W. Dür, J. I. Cirac, M. Lewenstein, and D. Bruß, Phys. Rev. A **61**, 062313 (2000).

[45] J. Eisert, PhD thesis, University of Potsdam, Germany, arXiv: quant-ph/0610253.

[46] H. J. Briegel, W. Dür, J. I. Cirac, and P. Zoller, Phys. Rev. Lett. **81**, 5932 (1998).

[47] D. Deutsch, A. Ekert, C. Macchiavello, S. Popescu, and A. Sanpera, Phys. Rev. Lett. **77**, 2818 (1996).

[48] M. Horodecki, P. Horodecki, and R. Horodecki, Phys. Rev. Lett. **84**, 4260 (2000).

[49] M. Horodecki, P. Horodecki, and R. Horodecki, Phys. Rev. Lett. **85**, 433 (2000).

[50] C. H. Bennett, D. P. DiVincenzo. J. A. Smolin, and W. K. Wootters, Phys. Rev. A **54**, 3824 (1996).

[51] S. Hill and W. K. Wootters, Phys. Rev. Lett. **78**, 5022 (1997).

[52] W. K. Wootters, Phys. Rev. Lett. **80**, 2245 (1998).

[53] K. Życzkowski, P. Horodecki, A. Sanpera, and M. Lewenstein, Phys. Rev. A **58**, 883 (1998).

[54] G. Vidal and R. F. Werner, Phys. Rev. A **65**, 032314 (2002).

[55] P. Rungta, PhD thesis, The University of New Mexico, USA (2002).

[56] H-P. Breuer, and F. Petruccione, *The Theory of Open Quantum Systems* (Oxford University Press, Oxford, 2002).

[57] M. O. Scully and M. S. Zubairy, *Quantum Optics* (Cambridge University Press, London, 1997).

[58] P. J. Dodd and J. J. Halliwel, Phys. Rev. A **69**, 052105 (2004).

[59] P. J. Dodd, Phys. Rev. A **69**, 052106 (2004).

[60] L. Diosi, *Irreversible Quantum Dynamics*, edited by F. Benatti and R. Floreanini (Springer, New York, 2003).

[61] T. Yu and J. H. Eberly, Phys. Rev. Lett. **93**, 140404 (2004).

[62] L. Jakóbczyk and A. Jamróz, Phys. Lett. A **333**, 35 (2004).

[63] T. Yu and J. H. Eberly, Opt. Commun. **264**, 393 (2006).

[64] M. P. Almeida, F. de Melo, M. Hor-Meyll, A. Salles, S. P. Walborn, P. H. Souto Ribeiro, and L. Davidovich, Science **316**, 579 (2007).

[65] A. Salles, F. de Melo, M. P. Almeida, M. Hor-Meyll, S. P. Walborn, P. H. Souto Ribeiro, and L. Davidovich, Phys. Rev. A **78**, 022322 (2008).

[66] J. Laurat, K. S. Choi, H. Deng, C. W. Chou, and H. J. Kimble, Phys. Rev. Lett. **99**, 180504 (2007).

[67] K. Kraus, *States, Effect, and Operations: Fundamental Notions in Quantum Theory* (Springer-Verlag, Berlin, 1983).

[68] T. Yu and J. H. Eberly, Phys. Rev. B **66**, 193306 (2002).

[69] T. Yu and J. H. Eberly, Phys. Rev. B **68**, 165322 (2003).

[70] M. Lucamarini, S. Paganelli, and S. Mancini, Phys. Rev. A **69**, 062308 (2004).

[71] D. Tolkonov, V. Privman, and P. K. Aravind, Phys. Rev. A **71**, 060308 (2005).

[72] Xian-Ting Liang, Phys. Lett. A **349**, 98 (2006).

[73] M. Yönaç, T. Yu, and J. H. Eberly, J. Phys. B **39**, S621 (2006).

[74] A. R. R. Carvalho, F. Mintert, S. Palzer, and A. Buchleitner, Eur. Phys. J. D **41**, 425 (2007).

[75] T. Yu and J. H. Eberly, *Quantum Information and Computation* **7**, 459 (2007).

[76] K. Roszak and P. Machnikowski, Phys. Rev. A **73**, 022313 (2006).

[77] V. S. Malinovsky and I. R. Sola, Phys. Rev. Lett. **96**, 050502 (2006).

[78] M. Ban, J. Phys. A **39**, 1927 (2006).

[79] M. Ban and F. Shibata, Phys. Lett. A **354**, 35 (2006).

[80] M. França Santos, P. Milman, L. Davidovich, and N. Zagury, Phys. Rev. A **73**, 040305(R) (2006).

[81] J. Wang, H. Batelaan, J. Podany, and A. F. Starace, J. Phys. B **39**, 4343 (2006).

[82] L. Lamata, J. León, and D. Salgado, Phys. Rev. A **73**, 052325 (2006).

[83] A. Jamróz, J. Phys. A **39**, 7727 (2006).

[84] T. Yu and J. H. Eberly, Phys. Rev. Lett. **97**, 140403 (2006).

[85] Z. Ficek and R. Tanaś, Phys. Rev. A **74**, 024304 (2006).

[86] G. S. Agarwal, *Quantum Statistical Properties of Spontaneous Emission and their Relation to Other Approaches* (Springer, Berlin, 1974).

[87] L. Derkacz and L. Jakóbczyk, Phys. Rev. A **74**, 032313 (2006).

[88] H. T. Cui, K. Li, and X. X. Yi, Phys. Lett. A **365**, 44 (2007).

[89] Z. Sun, X. Wang, and C. P. Sun, Phys. Rev A **75**, 062312 (2007).

[90] Jie-Hui Huang and Shi-Yao Zhu, Phys. Rev. A **76**, 062322 (2007).

[91] K. Ann and G. Jaeger, Phys. Rev. A **76**, 044101 (2007).

[92] M. Ali, A. R. P. Rau, and K. Ranade, arXiv:quant-ph/0710.2238.

[93] K. Ann and G. Jaeger, Phys. Lett. A **372**, 579 (2008).

[94] M. Ikram, Fu-li Li, and M. S. Zubairy, Phys. Rev. A **75**, 062336 (2007).

[95] M. O. T. Cunha, N. J. Phys. **9**, 237 (2007).

[96] J. H. Eberly and T. Yu, Science **316**, 555 (2007).

[97] F. Lastra, G. Romero, C. E. López, M. França Santos and J. C. Retamal, Phys. Rev A **75**, 062324 (2007).

[98] F. F. Fanchini and R. d. J. Napolitano, arXiv:quant-ph/0707.4092.

[99] G. Jaeger and K. Ann, Phys. Lett. A **372**, 2212 (2008).

[100] K. Ann and G. Jaeger, Phys. Lett. A **372**, 6853 (2008).

[101] A. G. Kofman and A. N. Korotkov, Phys. Rev. A **77**, 052329 (2008).

[102] A. R. P. Rau, M. Ali, and G. Alber, Eur. Phys. Lett. **82**, 40002 (2008).

[103] A. Al-Qasimi and D. F. V. James, Phys. Rev. A **77**, 012117 (2008).

[104] B. Bellomo, R. Lo Franco, and G. Compagno, Phys. Rev. A **77**, 032342 (2008).

[105] X. Cao and H. Zheng, Phys. Rev. A **77**, 022320 (2008).

[106] F. Lastra, G. Romero, C. E. López, N. Zagury, and J. C. Retamal, arXiv:quant-ph/0801.1664.

[107] M. Hernandez and M. Orszag, Phys. Rev. A **78**, 042114 (2008).

[108] I. Sainz and G. Björk, Phys. Rev. A **77**, 052307 (2008).

[109] I. Sainz and G. Björk, arXiv:quant-ph/0806.2102.

[110] R. Tahira, M. Ikram, T. Azim, and M. S. Zubairy, J. Phys. B **41**, 205501 (2008).

[111] L. Derkacz and L. Jakóbczyk, arXiv:quant-ph/0806.2537.

[112] J. Li, K. Chalapat, and G. S. Paraoanu, J. Low Temp. Phys. **153**, 294 (2008).

[113] Fa-Qiang Wang, Zhi-Ming Zhang, and Rui-Sheng Liang, Phys. Rev. A **78**, 062318 (2008).

[114] N. Yamamoto, H. I. Nurdin, M. R. James, and I. R. Petersen, Phys. Rev. A **78**, 042339 (2008).

[115] M. Scala, R. Migliore, and A. Messina, J. Phys. A **41**, 435304 (2008).

[116] C. J. Shan, W. W. Cheng, T. K Liu, J. B. Liu, and H. Wei, Chin. Phys. Lett. **25**, 3115 (2008).

[117] P. Li, Q. Zhang, and J. Q. You, Phys. Rev. A **79**, 014303 (2009).

[118] J. H. Cole, arXiv:quant-ph/0809.1746.

[119] S. Chan, M. D. Reid, and Z. Ficek, arXiv:quant-ph/0810.3050.

[120] J. Dajka and J. Luczka, Phys. Rev. A **77**, 062303 (2008).

[121] P. Marek, J. Lee, and M. S. Kim, Phys. Rev. A **77**, 032302 (2008).

[122] R. C. Drumond and M. O. Terra Cunha, arXiv:quant-ph/0809.4445.

[123] Y. Dubi and M. Ventra, Phys. Rev. A **79**, 012328 (2009).

[124] A. Al-Qasimi and D. F. V. James, arXiv:quant-ph/0810.0550.

[125] F. Lastra, S. Wallentowitz, M. Orszag, and M. Hernández, arXiv:quant-ph/0811.4399.

[126] J. P. Paz and A. J. Roncaglia, Phys. Rev. Lett. **100**, 220401 (2008).

[127] C. E. López, G. Romero, F. Lastra, E. Solano, and J. C. Retamal, Phys. Rev. Lett. **101**, 080503 (2008).

[128] L. Mazzola, S. Maniscalco, J. Piilo, K.-A. Suominen, and B. M. Garraway, arXiv:quant-ph/0812.3546.

[129] M. Ali, G. Alber, and A. R. P. Rau, J. Phys. B **42**, 025501 (2009).

[130] T. Yu and J. H. Eberly, Science **323**, 598 (2009).

[131] N. B. An and J. Kim, Phys. Rev. A **79**, 022303 (2009).

[132] K. Ann and G. Jaeger, arXiv:quant-ph/0903.0009.

[133] P. Meystre, M. O. Scully, and H. Walther, *Opt. Commun* **33**, 153 (1980).

[134] F. Shimizu, K. Shimizu, and H. Takauma, Phys. Rev. A **28**, 2248 (1983).

[135] T. Hellmuth, H. Walther, A. Zajonc, and W. Schleich, Phys. Rev. A **35**, 2532 (1987).

[136] A. Sanpera, R. Tarrach and G. Vidal, Phys. Rev. A **58**, 826 (1998).

[137] M. Ali, A. R. P. Rau, and G. Alber, in preparation.

[138] R. A. Horn and C. R. Johnson, *Matrix Analysis* (Cambridge University Press, Cambridge, 1985), Chap. 7.

[139] S. Bose, I. Fuentes-Guridi, P. L. Knight, and V. Vedral, Phys. Rev. Lett. **87**, 050401 (2001).

[140] J. S. Pratt, Phys. Rev. Lett. **93**, 237205 (2004).

[141] Z. Ficek and S. Swain, Phys. Rev. A **69**, 023401 (2004).

[142] D. P. Chi and S. Lee, J. Phys. A **36**, 11503 (2003).

Acknowledgments

The completion of a PhD and writing the thesis is usually the outcome of individual hard work and cooperation of several personalities. It was certainly not possible for me to accomplish this task without the roles of many individuals who have directly or indirectly influenced this work. First of all, I would like to submit thanks to my supervisor Prof. Dr. Gernot Alber. He has certainly played a key role in the completion of this thesis. He always encouraged me to think and work independently. His penetrating questions and criticism always opened the closed doors to me. I am indebted to him for accepting me in his group, for being patient, kind and always opening his doors to me.

I would also like to express my deepest regards to my collaborator Prof. Dr. A. R. P. Rau. In fact he introduced me to the main subject of this thesis. Almost all of my PhD work is in collaboration with him. I have benefited a lot from this collaboration. I am indeed thankful for his open mindness, encouragement, and always giving me many precious advises. In addition, I have learned professional ethics from him as well.

I am thankful to all present and past members of the *Theoretische Quantenphysik* group at TU Darmstadt for providing all sort of technical assistance to me. In particular, I thank Kedar Ranade and Oliver Kern who were not only always willing for discussions but they always gave their precious time and technical assistance whenever I needed it. In particular, I thank Oliver Kern for a nice drive to Innsbruck. I am thankful to Aeysha Khalique for her useful suggestions at some early stage of my PhD. I am thankful to Kedar Ranade for translating the abstract into German language and for reading this manuscript and pointing out several corrections. I am thankful to Ulrich Seyfarth for reviewing the German abstract. I am grateful to Sameer Varyani for providing a useful *Mathematica* code besides his valuable friendship.

I am thankful to Irmgard Praclik, Dietrich Praclik, Ute Stewart, Barbara Stowasser, and other members of IGM (International Generation Meeting) for exposing the German way of life and culture to me. Indeed I really enjoyed my stay in Germany due to these nice, kind and friendly people. I also appreciate ISO (International Students Organization) in this regard.

I acknowledge the financial support from Higher Education Commission (HEC), Pakistan and German Educational Exchange Service (DAAD).

I am unable to express my appreciation for my family members. All my brothers and sisters pushed me forward to achieve this goal. Indeed without their love, encouragement, taking care, moral and financial support at every stage of my life, I could never be able to reach this day. I am particularly indebted to my eldest brother Dr. Liaqat Ali and my dearest sister Ms. Farzana Anjum. I am thankful to my wife Qurat Ul Ann for her understanding and love. Although she entered in my life at the final stages of this work, nevertheless she managed to contribute by providing useful suggestions and Figures 3.1, 6.1, and 6.2.

Finally, I am thankful to Prof. Dr. R. Roth for being a referee for my thesis.

Die VDM Verlagsservicegesellschaft sucht für wissenschaftliche Verlage abgeschlossene und herausragende

Dissertationen, Habilitationen, Diplomarbeiten, Master Theses, Magisterarbeiten usw.

für die kostenlose Publikation als Fachbuch.

Sie verfügen über eine Arbeit, die hohen inhaltlichen und formalen Ansprüchen genügt, und haben Interesse an einer honorarvergüteten Publikation?

Dann senden Sie bitte erste Informationen über sich und Ihre Arbeit per Email an *info@vdm-vsg.de*.

Sie erhalten kurzfristig unser Feedback!

VDM Verlagsservicegesellschaft mbH
Dudweiler Landstr. 99   Telefon +49 681 3720 174
D - 66123 Saarbrücken   Fax   +49 681 3720 1749
www.vdm-vsg.de

Die VDM Verlagsservicegesellschaft mbH vertritt

Printed by Books on Demand GmbH, Norderstedt / Germany